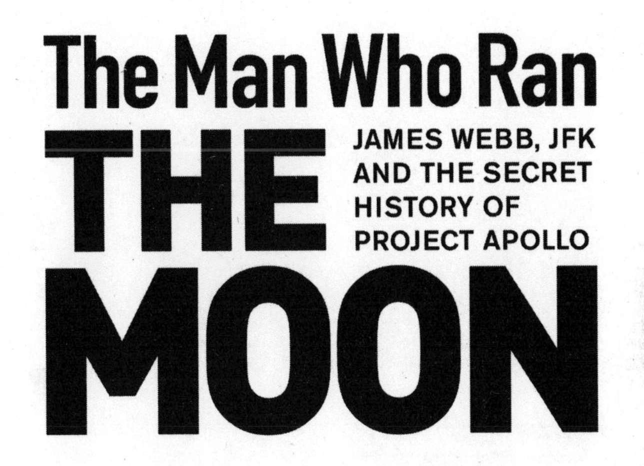
The Man Who Ran
THE
MOON
JAMES WEBB, JFK
AND THE SECRET
HISTORY OF
PROJECT APOLLO

PIERS BIZONY

The Man Who Ran THE MOON

JAMES WEBB, JFK AND THE SECRET HISTORY OF PROJECT APOLLO

ICON BOOKS

Published in the UK in 2006 by
Icon Books Ltd, The Old Dairy,
Brook Road, Thriplow,
Cambridge SG8 7RG
email: info@iconbooks.co.uk
www.iconbooks.co.uk

Sold in the UK, Europe, South Africa and Asia by
Faber & Faber Ltd, 3 Queen Square,
London WC1N 3AU
or their agents

Distributed in the UK, Europe, South Africa and Asia by
TBS Ltd, TBS Distribution Centre, Colchester Road
Frating Green, Colchester CO7 7DW

Published in Australia in 2006 by
Allen & Unwin Pty Ltd,
PO Box 8500, 83 Alexander Street,
Crows Nest, NSW 2065

ISBN-10: 1-84046-764-9
ISBN-13: 978-1840467-64-2

Typesetting by Hands Fotoset

Printed and bound in the UK by
Creative Print and Design Group

Contents

List of illustrations

About the author

Piers Bizony is a science journalist and space historian who writes for magazines such as *Focus* and *Wired*, as well as the *Independent*. His award-winning book on Kubrick's *2001: A Space Odyssey* was described as 'full of sparkling enthusiasm' by the *New Scientist* and 'excellent, in every way worthy of Kubrick's original precision-crafted vision' by the *Evening Standard*.

Foreword

Most people would say that President John F. Kennedy championed America's space agency NASA, but this isn't strictly correct. He didn't care much about space, and saw rocket exploration as little more than a Cold War contingency. It was a smart operator from North Carolina, James E. Webb, who steered the expansion of NASA from a minor collection of research labs into one of the grandest enterprises the world has ever seen. Aided by a cadre of Southern-born businessmen and politicians, Webb established a colossal network of influence in the service of space. For all its recent setbacks, NASA still commands influence today because of the legacy he created. He knew how to persuade not one, but two presidents in succession to give him what he wanted. This book is the first popular account of the man who ensured that America won the space race. No NASA administrator in the last 40 years has achieved half as much. Contrary to popular myth, it wasn't the German rocket scientist Wernher von Braun who led Apollo to the moon. It was Jim Webb.

We think of space administrators as dull, colourless bureaucrats. Webb was nothing of the kind. He was a powerful personality, combative, manipulative and driven. You underestimated him at your peril. He was a big, stocky bruiser of a man. When he walked into a room, people knew about it. His broad North Carolina accent and verbose way of speaking sometimes made him come across like a good ol' boy, a 'blabbermouth', as Bobby Kennedy once described him. Yet under all the down-home bluster, one of the sharpest and shrewdest political minds in American history was at work. He took hold of the space age and ran it just the way he wanted.

Webb knew everyone in Washington, and pretty much everyone in the business world. He understood what they wanted, and where his interests and theirs might converge. During his leadership of the space effort in the Kennedy and Lyndon Johnson years, he dispensed largesse, called in favours and occasionally strong-armed people who were foolish enough to oppose him. He was not a man to take 'no' for an answer. His enemies believed he was on a personal crusade to gain power and influence through his running of the space agency. And indeed he was. He was a visionary 'technocrat' whose ambitions stretched far beyond merely landing a man on the moon. By 1964 he held sway over 5 per cent of the entire US federal budget. The potential of that great wealth on the ground was just as important to him as the missions it could pay for in space. Here was a dyed-in-the-wool Democrat idealist who believed in personal honour, moral duty and national responsibility. He had come of age in the Depression years, and worked with Roosevelt during the New Deal era of national reconstruction. He believed passionately in the benefits of good government. And when crisis and scandal nearly destroyed the space agency he had championed, Webb wasn't afraid to shoulder personal responsibility, even as the press, and much of the political establishment, tried to destroy his reputation.

I'm not about to whitewash the contradictions and ambiguities of Webb's career. His success had its price, and there were compelling reasons why his name vanished from the glory roster of history for some years. In 1967, a launch-pad fire killed three astronauts. An investigation exposed some of the darkest secrets of the space age, including political corruption and questionable dealings among the powerful corporations to whom Webb had awarded major contracts for space hardware. The pure-hearted ideals of his plans were accompanied, in part at least, by back-room deal-making and cosy arrangements that veered perilously close to the unacceptable. Webb wasn't a dishonest man, but it's fair to say that he didn't like the way he ran NASA to be scrutinised too closely by outsiders. During 1967 he inadvertently came under the spotlight, giving journalists and Congressional representatives the appearance of a compromised and somewhat shady man. He and the great space agency he had built up came close to failure, and the ensuing scandal, largely forgotten today, broke his morale. His dream of a 'perfect' technological organisation for modern times nearly fell apart, and the bond of trust between him and his closest allies was strained to breaking point.

It took a few years, but Webb's reputation was eventually restored. His achievements have been properly acknowledged by Washington insiders, and by NASA too. Due for launch in about five years' time, the replacement for the popular Hubble Space Telescope will be called JWST: the James Webb Space Telescope. And as America reaches once more, perhaps, for the moon at the urging of President George W. Bush, Webb's determined style of management has much to tell us about how we should run space affairs today, and what new balances we might need to strike between strong, creative leadership and proper public accountability.

Webb is largely unknown to the general public. I hope this modest book goes some way towards restoring him to his rightful place as one of the greatest figures in the history of American government. I wrote this story because I was so disappointed that, among the many hundreds of mass-market books written about the 1960s NASA space effort, there were no popular biographies of the man who was *in charge* of the whole enterprise. Monumental biographies of Kennedy and Johnson barely mention Webb, despite the significance of space policy in the careers of both presidents. Many general histories of NASA and the space age, excellent though they are, refer to Webb only occasionally as a distant figure passing down commandments. We are told that he was a strong and capable leader, but very few texts explain why this should have been so. We gain no sense of him as a *person*. The astronaut biographies scarcely mention him at all. He can be retrieved piecemeal from the many exhaustive academic histories of NASA's bureaucracy if you look hard enough. I didn't quite find what I wanted in any of these materials, fascinating though they are. I was looking for the flavour, the tenor, the *voice* of the man himself.

NASA history is dominated by technical papers, engineering specifications and memos: so many memos. These mountains of paper tell us everything we could ever want to know about the decisions that shaped the space age. And at the same time, they tell us very little. As Secretary of State Dean Acheson so famously remarked: 'A memorandum is written not to inform the reader but to protect the writer.' Academic histories rely too much on memos because they are so satisfyingly formal, so solidly 'written in stone' into the historical record. This book is more concerned with the blurred edges of human relationships and personalities, and with the encounters in corridors and on windswept tarmacs, or the hasty telephone conversations or the dramatic meetings in conference rooms that spurred those memos to be written. The first time Jim Webb heard about the crisis that nearly toppled his agency, or learned of a plan to send men to the moon somewhat ahead of the expected timetable, it was on the telephone. No memos exist to relate the worst accusations laid against Webb by certain senators, because those charges were made during a 'closed executive session'. No memo, no history ... Except that *people* remember things. So long as they can talk, the history of key incidents survives, even if the memos are missing or were never written.

My unfortunate problem is that James Webb died in 1992. Even if he were still alive today, he would be a century old. Likewise, all too many of his co-workers have long since died. The age of Apollo was a frighteningly long time ago. My good fortune is that the voices of these remarkable people still speak. In the late 1960s and mid-1980s, a number of oral history projects were compiled by NASA-funded historians, by the Smithsonian Institution, and also by the John F. Kennedy and Lyndon B. Johnson libraries. Dozens of senior NASA staffers contributed to these taping sessions. The timing of the oral taping sessions was perfect. Interviewees are candid in their recollections of difficult times at NASA (and about each other) because they were safely retired at the time the tapes were made. On the other hand, with the tapes now twenty years old and more, we benefit from memories of events that were still relatively fresh in the interviewees' minds at the time those tapes were made.

I am extremely grateful to the Smithsonian National Air and Space Museum (NASM) Oral History project and its archivists and interviewers, most especially Martin Collins, David DeVorkin, Linda Ezell, Allan Needell and Joseph Tatarewicz. This book would not have been possible without their prior achievements, completed over a number of years. The NASM's chief space historian Roger Launius gave me much valuable advice and contributed an interview on his own account.

Likewise I was privileged to receive support and insight from Dr

John Logsdon, head of the Space Policy Institute at George Washington University. At the Lyndon Johnson Library, T.H. Baker drew fascinating stories out of Mr Webb. My thanks, also, to the WAMU radio station, whose insightful 1999 documentary about the Apollo project contained many useful inspirations.

In the NASA History Office, Steven Dick and his colleagues Stephen Garber, Jane Odom, Nadine Andreassen and Colin Fries were tirelessly helpful, sometimes providing me, unprompted, with materials critical to NASA which they felt were needed for my story.

Powering Apollo: James E. Webb of NASA is a fine scholarly book written a decade ago by the political science professor W. Henry Lambright. I would be lying if I said I hadn't drawn insights from it. I am also grateful to Charles Murray and Catherine Bly Cox, whose wonderfully engaging 1989 book *Apollo: The Race to the Moon* has become a standard touchstone for anyone interested in the lunar adventure. Charles very kindly sent me some materials and interview transcripts from his twenty-year-old files, along with notes made by the *Time-Life* reporter Robert Sherrod in 1969. I also acknowledge the investigative journalism conducted by the late Hugo Young and his colleagues more than three decades ago, while working with the London *Sunday Times* newspaper.

Above all, I have been privileged to talk directly with one man who worked immediately alongside Webb during the build-up to Apollo. Dr Robert Seamans, Associate Administrator of NASA from 1960 to 1965 and Deputy Administrator from 1965 to 1968, was a lively and open interviewee, and he also pointed me towards a mass of supporting documentation and useful books (including his own memoir, *Aiming at Targets*).

One last note: the Apollo adventure tends to be remembered as though it were an unblemished fairy-tale of nobility and goodheartedness. Mainly it was, and this writer is as much in thrall as anyone to the wonder of space flight and the magnificence of NASA's achievements, both then and now. Yet it's always worth remembering that space exploration is a human enterprise and not the act of angels or gods. I grew a little fond of the adventurers and opportunists who lurked at the edges of Apollo, despite their misdeeds. This book, in part at least, is their story too.

Part One
Accepting the Challenge

For all its power and influence, America is so often taken by surprise when great events shake the world. It was drowsy until Pearl Harbor; the CIA's multi-billion-dollar operation was caught by surprise when the Soviet Union collapsed almost overnight; and the World Trade Center attack was heralded by serious omens that weren't properly investigated until too late. America's mission to the moon was a similarly reactive, rather than proactive, phenomenon. On 4 October 1957, politicians and the press were alarmed at the news that Russia had launched Sputnik, the world's first artificial satellite. It seemed as if President Dwight D. Eisenhower was the only man in the country trying to remain calm.

Eisenhower held the White House on behalf of the Republicans, but Senator Lyndon Johnson controlled the Democrat majority in Congress. On that historic October evening, Johnson had been holding court at his beloved Texas ranch. After hearing the news about Sputnik, he took an after-dinner stroll to think things over from the good ol' cowboy vantage point he often liked to adopt. 'In the open West, you learn to live closely with the sky. It is a part of your life. But now, somehow, in some new way, the sky seemed almost alien.' The energetic, vociferous Johnson wasn't alone in his instinctive response that, from now on, space had somehow to be a part of the nation's security and prestige. Most Americans agreed with him that 'first in space means first in everything'.

A month after the launch of Sputnik, Eisenhower announced to the nation that he was appointing the first presidential science advisor. James R. Killian, the director of the Massachusetts Institute of Technology (MIT), had already won respect for his analyses of the relative balances of military power between the US and Soviet Russia:

trying, as it were, to differentiate serious facts from alarmist fictions. He now took a two-year leave of absence from his post at MIT and came to work for Eisenhower. Under Killian's chairmanship, the Presidential Science Advisory Committee (PSAC) gathered together various pre-existing strands of advisorship and brought them directly under the White House's supervision. Eisenhower sensed terrible dangers ahead. He wanted to maintain some sense of perspective in what threatened to become a heedless and expensive Cold War technology race, both on the ground and in space.

The PSAC was instructed to come up with a technical briefing about space. In a bold move that won many admirers, Killian decided instead to create a plain-language document that any ordinary person could read and understand. *Introduction to Outer Space* was released to the general public and became a bestseller. It explained how satellites get into orbit and what makes them stay up. It outlined America's possible future in space, but without setting any hard and fast dates. Above all, it tried to calm everyone's fears about the military aspects of rocketry. While stressing the practical value of satellites for reconnaissance and communication, *Introduction to Outer Space* stated flatly:

> *A satellite cannot simply drop a bomb. An object released from a satellite doesn't fall, so there is no special advantage in being over the target ... A better scheme is to give the weapon to be launched a small push ... But that means launching it from a moving platform halfway around the world [from your intended target] ... in short, the earth would appear to be, after all, the best weapons carrier.*

It was a cogent argument, and one that, half a century later, many people might still be wise to heed. However, the armed services, and all their related manufacturers, were lobbying vigorously for control of space. The Army believed that the main purpose of a nuclear ballistic missile was to launch from the ground and then hit an enemy target on another piece of ground, even if a brief arc into space occurred along the way. Therefore rocket technology was its legitimate concern. The Air Force argued that the earth's atmosphere, over which it already held dominion, transitioned smoothly into airless space with no obvious boundary. Anything that flew into the sky was its business, regardless of how high the thing went. In fact the word 'aerospace', so familiar to us today, was a confection of 1950s USAF publicists seeking to blur the distinctions between air and space. Meanwhile, the Navy's position was that they were best prepared to operate ships inside which men might remain sealed for many days or weeks at a time. A spaceship was just a new kind of submarine; a spaceman in the vacuum differed only in technical details from a diver in the depths of the ocean.

Hungry manufacturing corporations dispatched lobbyists to assault the President's restraints. As Killian recalled in later years:

> *Repeatedly, I saw Ike angered by the excesses, both in text and in advertising, of the aerospace-electronics press, which advocated bigger and better weapons to meet an ever bigger and better Soviet threat that they conjured up. I remember the shrill, hard campaigns by a few corporate lobbyists in support of their companies ... The Sputnik panic was being used to support an orgy of technological fantasies.*

Matters came to a head on 6 December 1957. Perhaps unwisely, an under-prepared Naval Research Laboratory rocket called Vanguard had been chosen to launch America's first satellite, and on the big day it blew itself to pieces on the launch-pad. It was a global humiliation. KAPUTNIK! STAYPUTNIK! FLOPNIK!, the newspapers derided. Eisenhower turned now to the Army missile teams, and their brilliant but somewhat embarrassing principal asset, Wernher von Braun.

German-born von Braun is probably the only 'rocket scientist' whose name is familiar to everyone. Born into an aristocratic family, he was obsessed by space. When the Nazis came to power he persuaded Hitler that rockets would make effective weapons. He and his dedicated team developed the V-2 'Vengeance' rocket at Peenemünde, on the Baltic coast. During the war, a hellish underground factory was built at Nordhausen under SS command, where half-starved slave labourers assembled the rockets.

In the last months of the war, British, American and Russian intelligence teams scoured the devastated German heartland for any remains of the V-2. It was, after all, the world's first guided missile, and it would define the future balance of power. Competition among the teams was ruthless, even though they were supposed to be allies. Von Braun decided that the Americans were his best hope, because they would probably employ him. The British couldn't afford him and the Russians might shoot him. He staged a brilliant escape for himself and his closest colleagues under the noses of SS squads, who were by now indiscriminately killing 'disloyal' Germans.

There followed a decade of half-hearted V-2 tests in the White Sands desert of New Mexico. Von Braun was frustrated by the US government's lack of interest in him, so he became a public campaigner. In the late 1950s, Walt Disney promoted him as a trustworthy crusader for space. Eisenhower was reluctant to allow von Braun a chance to send a satellite into orbit, even though the rockets he had by now developed on behalf of the Army Ballistic Missile Agency (ABMA) were more than capable of doing the job. The prospect of anything going aloft atop a military machine designed by someone with a Nazi war record didn't appeal to the President. However, Vanguard's failure forced a rethink. On 31 January 1958, von Braun's Juno booster successfully launched Explorer 1, America's first satellite. Instruments designed by physicist James van Allen immediately made important measurements of radiation belts surrounding the earth. This very real scientific achievement made up, in part, for the showy but effectively vacuous victory of the Sputnik, which had contained little more than a 'bleep, bleep!' radio transmitter.

The Army chiefs now lobbied with the righteous zeal of victors for the ultimate command of space. Eisenhower and PSAC moved swiftly to undercut them. Undoubtedly, the Vanguard fiasco had demonstrated the need for proper coordination of America's various competing rocket efforts. Just as surely, Eisenhower wasn't going to grant that role to the military.

One letter in an acronym

The National Advisory Committee for Aeronautics (NACA) was established in 1917 after senior figures at the Smithsonian Institution had expressed concern that European countries engaged in the First World War might outstrip America's fragile early successes in aviation. This was, after all, the country of Wilbur and Orville Wright. NACA was charged with conducting research in aerodynamics. It started out with a team of a dozen people and a budget of around $5,000 a year. A quarter of a century later, with a new global war looming, NACA still had only $3 million to disburse, along with a relatively modest complement of around 500 people. On the plus side, the little outfit had gathered together some brilliant scientists and engineers, all with a great passion for flight. A wind tunnel facility in Hampton, Virginia (the Samuel P. Langley Laboratory) was renowned for its groundbreaking work. Some of the young engineers crawling in and out of that tunnel were destined to become father-figures to the space age. NACA eventually expanded to the point where it was intimately involved with high-performance experimental aircraft, including the Bell X-1 rocket plane which broke through the sound barrier for the first time in 1948, with the legendary Charles 'Chuck' Yeager at the controls. By the late 1950s, NACA had established a major high-speed flight test centre in the Californian Mojave desert, and work was under way on the X-15 rocket plane, a thrillingly brutish black dart capable of reaching sub-orbital space.

The X-planes relied on substantial funding and other support from the armed services, however. NACA's in-house budgets remained modest. The planes drew a good deal of attention, although a shroud of secrecy in those Cold War years inevitably delayed any public announcements about new hardware or the latest speed and altitude records. To the world at large, NACA itself seemed a low-profile, gentlemanly organisation, brilliantly academic but not especially glamorous. In 1957, this courtly reserve was personified in NACA's last chief, Hugh L. Dryden.

Born in Maryland in 1898, Dryden built his academic reputation on the study of fluid dynamics. By 1924, he was setting up the world's first laboratory experiments trying to discover how air interacted with wing-shaped solid objects at the speed of sound. In the Second World War he served on several technical groups advising the armed forces. As head of a Washington project for the National Defense Research Committee (NDRC), he led development of the country's first guided missile successfully used in combat, the radar-homing 'Bat'. In 1947, to no one's great surprise, he was appointed Director of NACA, where he championed development of the X-15. His technical acumen was outshone only by his personal modesty. He had none of the worldly vices and resentments so often associated with directors of laboratories or government institutions. His profound Methodist convictions informed every aspect of his life. He might have been somewhat too thoughtful and distant for some tastes, yet he was undoubtedly a man people could trust. He was to become the wise, quietly spoken father figure to America's young space programme.

Dryden was already a member of Killian's PSAC advisory board when the decision was made to bring all space projects not directly allied to national defence under NACA's control. NACA, in turn, would have to become a federal agency, answerable to the President and to Congress. Lyndon Johnson, firmly in charge of the Democrat majority on Capitol Hill, drafted much of the legislation. He arranged that senior political figures in a 'Space Council' would lend the fledgling new agency some serious political weight, so that equivalently tangible achievements could be expected. Eisenhower was worried that such a Council would invest the rocket cadets with too much power. Johnson persuaded him that everything would be fine so long as the President himself, with all his cautions and caveats about space, remained firmly in charge of the Council. Eisenhower was coming to the end of his term in office, and Johnson had his eye on bidding for the White House in 1960. He saw himself commanding the 'high frontier' in space as well as on earth. He didn't see, out of the corner of his eye, the young and charismatic Jack Kennedy coming along to snatch away his rightful inheritance.

As Senate Majority Leader, Johnson's power rivalled, and in some respects even surpassed, that of Eisenhower himself. The wily, strong-willed Texan convened a special Senate Committee on Aeronautical and Space Sciences (we'll just call it the 'Senate Space Committee' from now on) with himself as chairman. This was the first time that a new permanent 'standing' committee had been established on the Hill since 1946. Now, the point of such a committee is to provide oversight for its particular area of concern: in this case, the American space effort. Johnson knew perfectly well that committee members wouldn't wish to be seen presiding over some no-account backwater activity. NASA's prospects were brightened merely by the formation of this committee. It was a sign that space affairs were to be treated seriously within Congress.

The National Air and Space Administration (NASA) was signed into law in July 1958. Johnson was gratified by the lack of resistance, initially at least, from the Department of Defense, but he had deliberately sent briefing papers and draft copies of the relevant legislation for Pentagon officials to examine during the Easter recess, and it seemed they had registered nothing more than the insignificant changing of one letter in an acronym, from NACA to NASA. 'They must have whizzed the paperwork through there on a motorcycle!' Johnson crowed. The generals woke up in alarm when they eventually realised what was happening. Soon enough, the Army Ballistic Missile Agency's rocket programmes, and even von Braun himself, were to become a part of NASA. The Pentagon fought a rearguard action to protect its 'latitude to pursue those things [in space] that are clearly associated with defense objectives', but in essence the matter was settled. The greater majesty and power of America's space effort was destined for civilian hands.

Keith Glennan, director of the Case Institute of Technology in Ohio, and a loyal Republican, took the reins of NASA, although he was courteous enough to insist that he wouldn't accept the job unless Hugh Dryden agreed to serve as his deputy. Dryden was widely respected, and his quiet, subtle representations had given NACA its head start in bidding for space, but his modest demeanour prevented him from winning the leadership of NASA itself. Congress wanted a man who came across as more obviously vigorous, especially in front of the press. The fact that Dryden was clever, decent and able was not sufficient. Dryden brought with him a NACA team called the Space Task Group, under the brilliant leadership of Robert Gilruth. An approximate design for America's first manned space capsule, the Mercury, had already been conceived at Langley. NASA now had the resources to make Mercury fly.

The supposed distinction between civilian and military activity lay more in the spirit than the letter of NACA's acronymic conversion into NASA. Navy and Air Force pilots dominated in the selection process for the first astronauts, while NASA's early space launchers were derived from ballistic missiles – Redstones, Titans and Atlases – shorn of their deadly nuclear warheads but otherwise indistinguishable from their Cold War cousins. Much of the 'systems engineering' expertise for managing NASA's operations also came from within the military. The Pentagon's thwarted ambitions for space were not so easily brushed aside. Tensions between NASA and the Department of Defense would never entirely dissipate. Indeed, one day they would contribute to the creation of a flawed, dangerous mess of a spaceship, the shuttle. But for now at least, Eisenhower had prevented space from becoming an immediate playground for the warmongers. This, at any rate, was his fervent hope as he left office. As the Supreme Allied Commander in the Second World War, he knew from first-hand experience what could happen when men in uniform took control of a nation's economy. On 17 January 1961, as he prepared to cede power to a newly elected Democrat president with ideas very different from his own, he made an extraordinary farewell address to the nation, in which he reminded American citizens that a gigantic and permanent arms industry had been created since the war where none had existed previously, and it was threatening to change all of society:

> *In the councils of government, we must guard against the acquisition of unwarranted influence, whether sought or unsought, by the military-industrial complex. The potential for the disastrous*

> *rise of misplaced power exists and will persist. We must never let the weight of this combination endanger our liberties or democratic processes.*

In a sideswipe undoubtedly aimed at the space cadets and other big spenders on federally funded research and development projects, Eisenhower also suggested that 'public policy could become the captive of a scientific-technological elite'. He would have been dismayed if he'd known that the newly formed NASA would soon be under the control of a man who believed that the 'scientific-technological elite' wasn't by any means a threat to America's future. It was the solution to all its ills.

One man from North Carolina ...

James Edwin Webb was born in October 1906, and raised in the small town of Tally Ho, North Carolina. He was the son of John Frederick and Sarah Gorham Webb. His father was superintendent of schools in Granville County. As a young man, Webb's first jobs were classic small-town means of helping his family to get along: working in farms, helping out in nickel-and-dime stores, and truck driving on construction sites. His father's role in education, and his mother's constant encouragement, exposed this highly intelligent young man to social and political ideas, essentially Democrat in nature. This environment, along with the habitually progressive instincts of his family, influenced him throughout his life.

After the Wall Street Crash of 1929, career opportunities for a young man in a small North Carolina town were limited, so Webb joined the Marine Corps. A handful of young men were to be selected as part of the first aviation squad in the Corps. This team was to be based in Long Island, New York. Webb's squad consisted mainly of privileged Ivy-Leaguers, yet he was readily accepted into the camaraderie of the group. He learned that he had the social skills to blend into any stratum of society and that he could match or even outstrip the achievements of people more privileged than himself. 'That was the first place, I think, that I realized I could compete with the wealthy fellows from Harvard and Princeton and so forth … I started at the very bottom, just getting by.' These experiences confirmed his Democratic view of America as a nation where opportunities should be allowed to overcome class and wealth. In the meantime, he knew it could do no harm to better his material career prospects. He chose a classic route: studying law.

On a trip to Washington to catch up with education officials he had known back in North Carolina, Webb was introduced to Congressman Edward Pou, who just happened to be in need of a new secretary. This was the era of Roosevelt's New Deal, a bold attempt to get America moving again after the Depression years, using vast sums of tax money to initiate new building projects and infrastructure development. 'I saw Roosevelt inaugurated, and then I saw all these tremendous things done in that famous first Hundred Days of his administration.' He then served as assistant to Max Gardner, a former Governor of North Carolina and now a powerful Washington attorney.

Webb turned out to be a smart political operator. In 1934, a number of Washington insiders had cause to take notice of this young man, not yet 30. A major scandal was threatening America's airmail services. Government contracts had been awarded in a less than savoury manner to a number of airlines. Roosevelt abruptly cancelled all the contracts and put the mail deliveries into the hands of the military. Overstretched and under-prepared pilots had to fly missions across hazardous mountains, on long-haul routes for which their training wasn't appropriate. Webb was called in 'to help get the airlines back on the job and stop these Army Air Force pilots from being killed every night'. As a man clearly known to be on the side of aviation *and* the Roosevelt administration, Webb was a valuable conduit in the negotiations. The airlines and their allied manufacturers created an Aeronautical Chamber of Commerce to represent their collective interests, and Webb helped ensure that Max Gardner, a close friend of Roosevelt's, was appointed as the Aeronautical Chamber's attorney. It was the first of many skilful power plays that Webb would make during his long career. The dispute was settled, and a new and less destructive relationship between government and the private airlines was established so that the postal air services could be resumed. Gardner's stock rose, of course, and he soon achieved his ambition to become a major lawyer at the heart of Washington's political community. As his influence grew, so did Webb's, and he joined Gardner's practice with Congressman Pou's blessing.

In 1936, Webb's experience in both law and aviation attracted the executives of the Sperry Corporation in New York. Over the next few years, as Sperry expanded operations for the Second World War, he came into contact with a large-scale high-technology environment for the first time. 'War is a hurly-burly kind of thing. I tried to learn all I could, but we had to move. I mean, we were hiring lots of people every day, training new people, getting rid of people who couldn't make the grade, worrying about the security problems, finding espionage agents sent over by Germany in our plant – just a million things to be done.' Happily married by now to Patsy Aiken Douglas after a five-year courtship, Webb helped shape Sperry into a successful defence contractor. As the workforce expanded from 800 to 30,000 people, he rose swiftly to the rank of vice president. Along the way, his legal skills saved the company millions of dollars of potential losses, and set the tone for his mastery of bureaucratic finesse.

The war was at its height. Defence contractors' warehouses were full of tanks, planes, ammunition and other equipment ready for urgent shipping to various battlefronts around the globe. So long as the government continued to make purchase orders, the war economy was sound and the contractors prospered. Webb knew this could be only a temporary state of affairs. One day soon, the war had to be concluded, both in Europe and in the Pacific. He urged Sperry to include a small but significant rider in their defence contracts. All equipment manufactured by Sperry as part of the war effort should be regarded as the property of the US government. By the time Webb departed in 1944 for a tour of duty with the Marine Corps as a training officer, the wisdom of this little paperwork shuffle was becoming clear. Sperry's rivals suddenly had vast amounts of weaponry in their stores that were surplus to requirements now the war was drawing to a close. Sperry's production runs were all safely paid for, regardless of whether or not the government now actually needed to ship their gear to the quietening battlefronts. Webb's reputation as a skilled corporate and political operator was now secure. 'These were very large contracts. If we had failed by something like a half of one percent to keep the inventory tied to the government contracts, then once the war ended, we'd have been out of business. We had a very small capitalization and were doing a very large business.'

Although he was never a technologist in his own right, Webb learned how to manage the relationships between men and complex machines under the most pressing conditions. On his return to active service in the Marines, he coordinated the use of radar control for night fighters supporting US troops in their Pacific island landing operations. The planes had to attack ground targets without accidentally bombing their own people. Radar equipment was modified so that ground troops could carry the components on their backs, then reassemble the sets on shore. All this required close and urgent liaison between soldiers, airmen, radio technicians and industrial contractors, few of whom spoke the same language or thought about problems in the same ways. The fact that Webb's younger brother Gorham was being held at that time in a Japanese POW camp lent an even greater urgency to his work. (Gorham survived his internment.)

After the war, Webb returned to public service, rising to the rank of Director of the Bureau of the Budget under President Truman, a position he held until 1949. He fought a running budget battle with the Navy chief William Forrestal, and the animosity between the two men was exacerbated when Forrestal was appointed as America's first overall Secretary of Defense in 1947. Webb's other challenge as budget director was the sudden emergence of the 'Marshall Plan', a gigantic aid package designed to prevent a war-scarred Western Europe from falling under Soviet domination.

Truman sympathised with Webb's efforts to keep Forrestal's military spending in check. In 1949, at the start of the President's second term, Webb was rewarded with the more senior post of Under Secretary of State. He immediately tried to impose management systems that would enable the State Department to wield more influence, and not become a de facto second arm of the new Department of Defense. He also became a passionate advocate of science and technology. Specifically, he ensured that major US embassies had science officers who would perhaps feed back an understanding of technology into the State Department's overall worldview. Webb's boss, Secretary of State Dean Acheson, was more concerned with the immediate problems of the Korean War. He and Webb never really saw eye to eye about priorities.

When the Truman administration ended in 1953, Webb was bruised and battered but immensely experienced. He had made some enemies, yet no one doubted the complexities of the issues he had tried to deal with, nor that he had scored many notable successes. Nevertheless, it was time to move on. He didn't like the idea of working for the incoming Republican president, Dwight D. Eisenhower.

Fortunately, a powerful senator and businessman, Robert 'Bob' Kerr, had a new job for him. 'Mr Webb, you're the kind of man we need in Oklahoma.'

Webb, who hadn't exactly thought of Oklahoma as his next opportunity in life, was nonplussed and said nothing.

'You're not listening to me', Kerr's booming voice drove on. 'I know you're not listening to me because you haven't asked me *how much* you're needed in Oklahoma. One million dollars is how much.'

... and one from Oklahoma

In the autumn of 1911, when Robert Samuel Kerr enrolled as a high school junior at Oklahoma Baptist University, the dean asked what he wanted for the future. Without hesitating, the tall, brash young man said: 'I want to be a very successful lawyer, make plenty of money, marry a beautiful and good girl, have children, and live in a first-class house.' These were not easy ambitions for an Oklahoman born into considerable poverty, but in the coming decades Kerr achieved all these dreams and more. He became an oil tycoon, the governor of his home state, and eventually a senator. But he suffered some daunting setbacks along the way. In 1921 his twin daughters died at birth, and his first family business, a produce wholesalers, went up in flames. Three years later his wife and a son died in childbirth. Kerr was grief-stricken for many months and vowed never to remarry. In 1925, though, he met and wed Grayce Breene, the daughter of an oilman from Tulsa. Kerr was so short of cash by then, he had to borrow a thousand dollars from his new wife's family to launch the marriage. Little wonder that he once said: 'I'm a wife-made man and I suspect there actually are more wife-made men than self-made men.'

By the end of the 1920s, Kerr had teamed up with his brother-in-law to form the flourishing Anderson and Kerr Drilling Company. Anderson ran much of the practical side of the operation while Kerr suavely finessed the investment capital. Following Anderson's retirement in 1937, Kerr joined forces with another successful oilman, Dean McGee, and the Kerr-McGee company began to take shape. It would become a cornerstone of Oklahoma's economy, with interests ranging from oil and gas to banking, real estate and nuclear energy. Its influences would stretch into the space age and beyond. From now on, it was wise for anyone seeking to do business in

Oklahoma to consider how Kerr's interests might best benefit from theirs.

Kerr rose to political prominence in the 1940s. His critics rounded on him as a 'special interests' politician. He remained entirely unapologetic throughout his life. 'Whether there's any oil in the future of politics, there's going to be a lot of politics in the future of oil.' Time has certainly proved the validity of his instincts. Shameless in championing his personal agendas, he often said: 'I represent myself first, Oklahoma second, and the United States third, and don't you forget it!' Of course he wasn't so totally cynical. He was popular in his home state, not least because he achieved much of real substance to improve the lives and job prospects of his constituents; but behind the comedy routines and the larger-than-life charm, he was indeed a great lover of power and influence.

As Jim Webb was contemplating his future in the spring of 1953, Kerr was looking for someone to turn around the fortunes of a Kerr-McGee subsidiary, the Republic Supply Company. This outfit was the main source of all the pipes, rigs and other physical equipment that Kerr-McGee needed to drill for oil, and it was in a mess. Webb decided to come aboard, fix the Republic problems, make that one million dollars' prize money and give himself some financial stability for the future. He made quite a success of his time in Oklahoma. 'I was sort of looked at askance as a newcomer for a while, and then at some point people began to say, "He's running his business, he's paying his debts at the bank and so forth." The banks knew I had taken over a badly-run company [Republic Supply] to rehabilitate, and the only way I could do it was to expand our sales.'

Nevertheless, Webb never promised Kerr that his stay would be permanent. After five years, he asked: 'Have you made money on me? Are you satisfied?' Kerr said he was delighted at his performance, and wanted him to sign another five-year contract. Webb politely turned him down. 'I could see that he wanted to get a collar around my neck, and I didn't want that.' Kerr appeared to accept his friend's desire for freedom. In truth, the wily Oklahoman was busy stitching a new collar.

In 1961, Lyndon Johnson's acceptance of the Vice-Presidency required that he surrender most of his old territory on the Hill. He gladly handed over the Senatorial reins of space to his friend and ally Kerr. He, in turn, lost no time considering how this prestigious appointment might benefit his interests. Political journalists could only look on in wonder. 'In the Congressional wheeling and dealing to land juicy contracts for their home states, all members are equal, but some are more equal than others. And Robert Samuel Kerr of Oklahoma, chairman of the Senate Aeronautical and Space Sciences Committee, is the most equal of them all.' NASA was about to become his favourite new game, and his erstwhile Kerr-McGee vice president, Jim Webb, was the first piece he moved across the board.

Recruitment

Since July 1958, the National Aeronautics and Space Administration's first chief, Keith Glennan, had led a small research agency whose task was to launch tiny satellites and develop simple Mercury capsules to go atop low-powered missiles, in order to keep up (barely) with Russian developments in space. Eisenhower had insisted that Mercury was not to be a long-term exercise, and Glennan readily agreed. By his own admission, he didn't believe in the future of the project he was running. 'I was never a space cadet', he admitted in later years. NASA was nothing like the gigantic organisation we have become accustomed to. When Kennedy came to power in 1961 the Mercury project was undergoing its first test flights, but no astronauts had yet flown. Eisenhower had sanctioned modest spending on future space projects research, orbiting space stations, trips to the moon and so forth, all meticulously realised on paper, yet no one seriously thought these things could be done in reality any time soon. In the 1960 election campaign, Kennedy had taunted Nixon about Eisenhower's perceived lack of zeal in defence spending. He exploited rumours about Soviet weapons and their supposedly daunting array of nuclear-tipped ballistic weapons (the so-called 'missile gap'). Other than that, he had taken little interest in rockets of any kind. Mercury was widely considered by NASA enthusiasts as the opening gambit in the long-range conquest of space. Many Washington insiders viewed 'man-in-a-can' as a one-off deal, a necessary circus stunt that would be over and done with soon. They'd get an American in space in the next few months and put the shame of Sputnik behind them once and for all. And then all that crazy Buck Rogers nonsense would quieten down to a more measured pace.

Glennan was more than happy to get back to his beloved Case

Institute as soon as possible. Kennedy and Johnson had to find someone – a Democrat, obviously – to take over NASA and see it safely through the Mercury programme. The trouble was, no one of any merit wanted the job. Despite popular fascination with the astronauts, and the obvious short-term glamour of the thing, space didn't look as if it had any future. NASA's relatively modest stature in federal circles at that time was discouraging for anyone who truly fancied his chances in the Washington rat race. The Army and the Air Force had to be kept at bay, and worst of all, there was the risk of losing an astronaut atop one of those volatile rockets, and subsequently having to accept responsibility as administrator. At least seventeen people turned down the White House's requests to become the new chief of NASA.

And then there was the problem of the Vice President, Lyndon Baines Johnson. Anyone who took on space would also have to deal with LBJ and his tantrums and his wheedling and bullying and all the rest of it. This, too, was not a happy prospect for anyone who'd gone through the 'Johnson treatment' and lived to tell the tale. Those seventeen refusals made him look bad. They enhanced the mocking portrait some political commentators had drawn of a man whose days as a true power in the land were behind him: an ex-heavyweight Senate Majority Leader reduced to nothing more than a toothless Vice-Presidency. Much has been written about the backroom deals between the Kennedy brothers and Johnson, in which he was persuaded to sacrifice his presidential ambitions in favour of supporting Kennedy. Once installed as Vice President, he was assigned leadership of the Space Council: partly because he wanted it, but rather more because Kennedy hoped to give this noisy and troublesome Texan something useful to do – something suitably glamorous, and at the same time fairly trivial and harmless, which Kennedy didn't think he'd need to care about personally.

Senator Kerr now made his move. As Webb understood events: 'Johnson was called down to Palm Beach several times [on campaign business]. During one of those trips, he and Kerr were both together. Apparently they talked about whether I should go into the Treasury.' Kerr then put in a phone call to Webb, but he really wasn't interested. He'd served his time in years past, and that was enough. He had thought the matter was closed, but on the morning of Monday, 30 January 1961 he was summoned to Johnson's office, after just a weekend's advance notice to think things through. There'd been a telephone message on Friday from Jerome Wiesner, ex-PSAC member and now Kennedy's science advisor, asking Webb to see the Vice President about running NASA. Filled with apprehension, Webb flew into Washington and talked with a number of friends and trusted associates over the weekend about what he should do. He confirmed what he'd already sensed: Wiesner and Kennedy weren't much interested in space, and NASA's low-key progress so far hadn't stirred many hearts among the Camelot crowd.

Much to Webb's relief, when he arrived at Johnson's office for his appointment, he found NASA Deputy Administrator Hugh Dryden sitting quietly on the sofa. Johnson had wanted him to meet Webb and help persuade him into leading the Agency. Webb lost no time telling Dryden: 'I don't think I'm the right person for this job. I'm not an engineer and I've never seen a rocket fly.'

'I agree. I don't think you are either', said Dryden.

'Well, can you tell the Vice President?'

'I don't believe he wants to listen to me on that.'

Just at that moment, Secretary of the Army Frank Pace, a Webb ally of many years' standing, arrived on other business. One of Pace's current projects was the construction of the Atlas intercontinental ballistic missile (ICBM). He, too, agreed that rockets probably weren't Webb's field. 'How about you telling the Vice President?' Webb pleaded. 'You had an appointment first. We'll just step out, and when the Vice President comes in you can talk to him.' Pace said he'd try. When Johnson was ready, Webb and Dryden stepped out to an adjoining room. After a discouragingly short time, the door to Johnson's office flew open and Pace flew out of it, obviously shaken. The Vice President wasn't in the mood for another turndown, and Pace had swallowed a dose of the 'Johnson treatment' on Webb's behalf. There was a last, quick moment to chat with Phil Graham, publisher of the *Washington Post*, who'd just arrived on press business. Then it was Webb's turn in the lion's den. Johnson made it quite clear that he and President Kennedy wanted him to run NASA. On his way out, Webb tried his chances with Graham, who told him the only person who could get him out of his predicament was Clark Clifford, a Kennedy advisor. A helpful Johnson staffer took him to another office across town, from where he made a frantic phone call to Clifford, who simply laughed. 'I've been recommending you! I am not going to get you out of it!'

One argument Webb couldn't use to escape the job was the inconvenience of moving back to Washington. He had already relocated his family from Oklahoma so that he could more effectively serve as chairman of the Municipal Manpower Commission, a Washington-based body funded by the Ford and Sloan Foundations. 'We were looking at the problem of personnel requirements to keep our American cities from deteriorating.' Continuing business interests in Oklahoma and elsewhere kept Webb fully occupied. He clearly wanted to maintain his influence at the heart of Washington life, but 'I was beginning to back away from the formalized, general manager type of job, and spending more time on public service, like Educational Services, the Municipal Manpower Commission, and several other advisory committees on which I served.' Although he was also on the Board of Directors for the McDonnell Aircraft Company, manufacturers of the Mercury capsule, his experience in rocket technology was limited. The last thing he had in mind was to take control of America's overall space effort. He decided to lever himself some political capital. 'I made it clear to the Vice President that I would not accept the position except by the direct invitation of the President. I wouldn't take a second-hand invitation to it.'

Johnson organised for Webb to go over to the White House and see the President immediately; whereupon Webb again reiterated his belief that he wasn't the right man for the job. 'You need a scientist or an engineer', he insisted.

Kennedy knew all the right buttons to press. 'No, I need somebody who understands policy. You've been Under Secretary of State, and Director of the Budget. This program involves great issues of national and international policy, and that's why I want you to do it.'

'Well, what do you want of me?' Webb asked. 'Do you want me to find out what needs to be done, and then do it? Or am I expected to come in here and try out some preconceived pattern?'

Kennedy assured him that his job really was to lead NASA and not merely to rubber-stamp other people's existing ideas about space. Webb accepted the post, although he always insisted: 'I would not have taken the job if I could honorably and properly not have taken it.' No reasonably ambitious man would want to turn down a chance to do important work for his country. His reasons for demanding a direct presidential request were, of course, subtler than mere vanity. He wanted to leave the White House that day armed with the fullest authority he could muster. The Kennedy administration now owed him a debt of gratitude for taking a job they couldn't sell to anyone else in town. He would make sure he collected on it. As for the Johnson problem: Webb may have felt somewhat coerced into the NASA job, but he was grateful that at least Kerr had given him some covering fire. 'He told Johnson this story about my independence – how I wouldn't be kicked around – which meant Johnson never tried.'

There's no reason to suppose that Kennedy and his team were especially enamoured of Webb. If they had known how important space was about to become, they would undoubtedly have appointed someone more to their liking. As the political writer Charles Murray observed:

> *Stocky and voluble, Webb at fifty-five was of a different generation than most of the others in this new administration, and from a different world. Instead of Harvard and wire-rimmed glasses, clipped accents and dry wit, he was University of North Carolina and rumpled collars, corn-pone accent and down-home homilies, a good ol' boy with a law degree.*

Picture him in his mid-fifties, vigorous, noisily garrulous and with a pronounced North Carolina accent. Imagine the slight twinge of unease among the East Coasters of Virginia and Boston when he talked about the Mercury 'cap-soole'. Former NASA chief historian Roger Launius describes Webb as a man with a pronounced drawl: 'He'd play that good ol' Southern boy "I just fell off the turnip truck" routine, and he was a master at it. By the time you were finished talking to him, you had agreed to all kinds of things that you'd never conceived of before.' Picture, also, the respect engendered by his physical presence: though not a tall man, his strong, square head and bullish neck, his sturdy chest, an obstinate jaw and narrowed grey-blue eyes lent him a dominant demeanour. He kept his peppery hair cut short at the sides and wore his suits tidily rather than fashionably. He didn't need slick tailors to boost his image.

And if Lyndon Johnson had his 'treatment' to dole out, Webb employed similar techniques, albeit with greater charm and considerably less reliance on barnyard insults. (He was polite in all his dealings, although his courtesies could sound icy when he wanted them to.) His blizzard of words could intimidate, confuse or impress, depending on his audience. One observer said that having a conversation with him was 'like trying to get information out of a fire hydrant'. John F. Kennedy's brother Bobby underestimated Webb's subtlety, thinking him nothing but 'a blabbermouth'. He couldn't have been more wrong, for Webb was widely acknowledged as one of the canniest political operators ever to hit Washington. It was best for anyone going up against him not to be fooled by all the babble and noise. One time, a friend who regularly dined with Webb bet with his secretary that he'd be able to get through an entire lunch without saying a word. Webb would take up all the slack without noticing. Sure enough, the lunch in question came to an end, and the friend spoke hardly at all apart from grunting a brief 'hello' and 'goodbye' at the appropriate times. Webb got up from his chair and said: 'That was one of the best lunches we've had together. I learned so much.'

And he probably *had* learned something, for he was a man who liked to think aloud. He invited his listeners to interject or to act as sounding boards. He did listen in a more focused way, of course, to those whom he respected. Within NASA, the quiet-spoken counsel of Hugh Dryden was particularly valuable to him as he grappled with the relatively new and unfamiliar world of space engineering. On the other hand, if any particular voice was raised in opposition too often, Webb was seldom afraid to fight back. He was obstinate, combative and fiercely determined in everything he undertook. Not even a president of the United States could contradict him without being challenged.

The second and arguably greatest administrator in NASA's 50-year history began work on 14 February 1961, after insisting that he cleared all the formalities on the Hill first, including a confirmation hearing (albeit one chaired by the amenable Senator Kerr). 'The town was full of young men who were riding around in government cars, and already had their offices set up, and they hadn't even been confirmed yet.' Webb, a lifelong student of political niceties, knew better than anyone how 'little things like that can break you in this town'. Before his tenure at NASA was ended, he would experience the bitter truth of his instincts.

Taking the reins

Now the problem became: how much of NASA's current legacy from the Eisenhower administration did Webb want to live with? Certainly he wanted to keep Hugh Dryden, who had been passed over yet again for the top post because, as usual, Johnson, Wiesner and others felt he was too diffident. Webb respected Dryden, and insisted that he be sworn in as Deputy Administrator.

Next on the list of variables was Robert Seamans, NASA's Associate Administrator, and third in command of the Agency up until that time. He had essentially been the 'general manager', although that term was not used as a formal title. Seaman's grasp on the countless everyday details would be important. On the other hand, if he and Webb had any profound differences in philosophy, that might be a problem. After all, Seamans was a committed Republican ...

Robert 'Bob' Seamans had been an aeronautical professor at MIT before working for the Radio Corporation of America (RCA) on airborne radar systems and missile electronics programmes. In June 1960, Keith Glennan decided that he and Dryden needed a third helper to impose management discipline on the Mercury capsule project. He called Seamans. 'We want you to get in there and take charge!' He did his best to comply, although at first he found NASA's layers of command and accountability to be somewhat shapeless and ill-defined. 'The way the organization was set up, it was difficult for me to exercise my responsibility. There was no place for me to grab hold of the total NASA program.'

Seamans, at 47 years of age, had a youthful flavour to him. Tall, sandy-haired, and accessorised with one of those pairs of glasses that give the impression of a comic-book superhero disguised as a high-school intellectual, he had an optimistic mindset and a tremendous grasp of managerial detail. Now he awaited his fate under a new

NASA chief. He recalls an amiable first meeting. Webb joked that at least a Republican would guarantee NASA some bipartisan support in Congress. They compared notes about various Washington figures: it was a short-hand way to find out if they had the same kind of instincts about the various personalities on the Hill that they'd have to deal with from now on. Seamans was new to the political swamp, so Webb changed tack and discussed several major American corporations and how they were structured. Happily enough, both men tended towards 'flat' organisations with relatively few layers of middle-management separating the upper echelons from the ground troops: an uncluttered system of ranking where individual responsibility and accountability was high, yet rewarded with an equivalent degree of personal autonomy and decision-making power. They veered away from complicated pyramid-shaped hierarchies that caused people to look upwards for authority on the smallest tasks before implementing them, and downwards for people to do the actual work. That was not a system designed to get the best out of individuals or, indeed, NASA.

Having established that their philosophies of management were not drastically at odds, Webb suggested that a 'Triad' be established, wherein he, Dryden and Seamans would effectively run NASA together, albeit with Webb having ultimate veto on all major decisions to do with the Agency as a whole. It made sense: Webb needed all the help he could get. Dryden was his sounding-board, while Seamans would literally manage and implement all those decisions which Webb wouldn't necessarily be able to attend to in fine detail, because *his* job was dealing with Congress and the President, all the while trying to control a swarm of private and corporate vested interests to keep them more or less on NASA's side. 'The benefit of the three of us was that we liked each other, we knew each other, we had great confidence in each other, and we didn't see any of the backbiting that occurred in so many organizations.' From now on, Webb would show an extraordinary degree of trust in his NASA associates, especially when it came to technical and engineering decisions about rocket programmes, which they, rather than he, were more qualified to make.

That trust was tested almost immediately upon Webb's appointment. In July 1960, the first unmanned test of a Mercury capsule had ended in disaster. Its carrier rocket was the new Air Force ICBM, the Atlas. An adapter section between the rocket and the capsule had failed, rather than the rocket itself. Now NASA's Mercury team was preparing another unmanned test flight with the Atlas, and the Air Force was greatly concerned. Webb was contacted, and 'urged not to make that flight, for the reason that if it failed, it would raise questions as to the credibility of our nuclear missile deterrent force. Now this presented a very hard question to me as Administrator. I got our people in and asked very thoroughly and carefully whether or not we had confidence we were doing the right thing and were ready to go. They all said "Yes." So I sent word back [to the Air Force] that we were ready to go.'

Whereupon the Air Force made representations to the White House that the flight shouldn't be allowed to happen. Webb thought this flaring of nostrils was mainly for show, a knee-jerk challenge to NASA's independence rather than an objection based on sound reasoning. 'We were like two strange animals who paw around and smell each other, testing each other out ... They were writing speeches and passing out stuff to the newspaper people saying that we couldn't possibly succeed. We were going to stumble and drop the basket of groceries, and they'd have to pick it up. And they were sort of sticking out their foot to trip us up.' White House science advisor Jerome Wiesner, the very man who'd helped recruit Webb in the first place, was more than happy to listen to the Air Force's complaints. He had already written a strongly worded report to Kennedy:

> *Because of our lag in the development of large boosters, it is very unlikely that we shall be first in placing a man into orbit around the earth ... A failure in our first attempt to place a man into orbit, resulting in the death of an astronaut, would create a situation of serious national embarrassment ... We should stop advertising Mercury as our major objective in space activities.*

In subsequent interviews Webb couldn't remember if he contacted Kennedy or the other way around, but inevitably the two men had to speak together to settle this important issue and get Wiesner and the Air Force generals off NASA's back. 'I said, "Now look. If you want me to run this program, I'll run it. And if you stick with me, I think we have some reasonable chance of coming through this thing together. If you want these other people to run the program, I don't know how you're going to come out." And that's basically the way things came to be settled. In each case, Kennedy said, "I'm going to stick with you."'

Just a week after starting his job at NASA, Webb's faith in his people was rewarded. An unmanned Mercury was launched on 21 February 1961 atop an Atlas missile for a successful test of the capsule's heat shield and parachute recovery systems. Next day, NASA announced that Alan Shepard, John Glenn and Virgil Grissom had been short-listed as candidates to make the first manned flight into space. This trip wasn't to be a full orbit. No one yet knew how a man might stand up to a prolonged period of weightlessness, and the Atlas missile was not yet cleared to fly a living cargo. One lucky astronaut would undergo a brief fifteen-minute sub-orbital cannon-shot atop a small but reliable Redstone rocket developed by von Braun and his old ABMA team, now working within NASA.

A last-minute snag had to be fixed before the mission could go ahead. Space historian John Logsdon takes up the story. 'The first American manned flight should have happened in March 1961, but a previous Mercury test with the Redstone on January 31 had a chimpanzee called "Ham" on board. The retro rockets fired late, sending Ham a hundred miles downrange of the correct splashdown zone, and it took several hours to recover him, which made for one very unhappy chimpanzee by the way. The technical problem was very simple, very easy to fix, but they had to do another test of the Mercury before committing a human. So it's an interesting question: what would have happened if we had made that original date? I think history would have worked out very differently.'

NASA's official log of the Mercury endeavour notes that Ham appeared to be in good physiological condition, but sometime later, when he was shown the spacecraft, it was visually apparent that he had no further interest in cooperating with the space flight programme. Ham certainly earned the long and comfortable retirement awaiting him in a special zoo. Electrical short-circuits during his flight and water leakages after splashdown played havoc with the Mercury's cabin, causing the chimpanzee considerable distress. No one wanted to take any chances with a capsule designed to send the first man up, in case he also ran into difficulties. So a slight delay entered the schedule. Meanwhile the press mocked NASA's preference for primates over pilots. A witty cartoon had a pair of apes casually strolling away from a just-landed space capsule. One turns to the other and says: 'I think we're behind the Russians but slightly ahead of the Americans.' Another cartoon showed an equally blasé simian pointing to a diagram on the wall, briefing his human colleagues on what to expect during a Mercury mission: '… Then, at 900,000 feet, you'll get the feeling that you *must* have a banana.'

Shooting for the moon

In his first weeks in office, the new president was most assuredly not showing the tremendous interest in space for which history tends to remember him. When Webb held his first serious budget discussions with Kennedy on 22 March, the NASA chief's feeling was that 'we had to go forward with the development work on big boosters [the Saturns], and we had to start some work toward the spacecraft to fly on them, following Mercury.' Kennedy said he'd approve the boosters, along with a weather satellite capable of providing hurricane alerts, but for the time being NASA's request for a complex multi-man spacecraft, known as Apollo, was put on hold. Mercury hadn't done much so far to impress the White House. In the absence of proof that spaceships with men aboard could deliver political advantages in return for their great expense, Kennedy preferred not to be rushed into approving more of them. His support for work on the Saturn was a holding pattern designed to keep his options open for the future. Somewhat like Eisenhower before him, he agreed that America must build large rockets – larger, of course, than anything the Russians could come up with – but he wasn't yet ready to think about the payloads they might carry.

His time for thinking ran out just three weeks later. The CIA and other intelligence agencies had been reporting for some while that the Soviets might be preparing to launch their own manned capsule. The rumours rapidly solidified into fact. On 11 April 1961, Kennedy appeared on an NBC early-evening television programme sponsored by Crest toothpaste. He and his wife Jacqueline talked with reporters Sander Vanocur and Ray Scherer about the difficulties of raising their small children, and about the President's 'hands-on' management style. Kennedy mentioned that political events often appeared more subtle and complex from inside the Oval Office than they did to the outside world. Even as he smiled and joked for the television cameras, he knew from intelligence reports that a significant defeat awaited him in just a few hours' time.

Russia's dawn was America's night. At 1:07 am Eastern Standard Time, NATO radar stations recorded the launch of a large R-7 Soviet rocket, and fifteen minutes later a radio monitoring post in the Aleutian Islands detected unmistakable signs of live dialogue with its human occupier. Wiesner called Kennedy's press secretary Pierre Salinger with the news. Kennedy had gone to bed a few hours earlier with a sense of foreboding. Wiesner asked him, did he want to be woken as soon as more information became available? 'No', the President answered wearily. 'Give me the news in the morning.'

At 5:30 am Washington time, the Moscow News radio channel announced the latest Soviet triumph in space. An alert journalist called NASA's launch centre at Cape Canaveral in Florida to ask: could America catch up? Press officer John 'Shorty' Powers was trying to grab a few hours' rest in his cramped office cot. He and many other NASA staffers were working sixteen-hour days in the lead-up to Mercury's first manned flight. When the phone at his side rang in the pre-dawn silence, he was irritable and unprepared. 'Hey, what is this!' he yelled into the phone. 'We're all asleep down here!' Later that morning, the headlines read: SOVIETS PUT MAN IN SPACE. SPOKESMAN SAYS US ASLEEP.

Yuri Gagarin, a 27-year-old Soviet Air Force pilot, had just become the first human ever to fly in space, aboard a small capsule named Vostok ('East').

Kennedy held an uncomfortable press conference. Normally a self-confident and eloquent public performer, he seemed distinctly less sure of himself than usual. He was asked: 'Mr President, a member of Congress today said he was tired of seeing the United States coming second to Russia in the space field. What is the prospect that we will catch up?'

'However tired anybody may be – and no one is more tired than I am – it is going to take some time. The news will be worse before it gets better. We are, I hope, going to go into other areas where we can be first, and which will bring perhaps more long-range benefits to mankind. But we are behind.'

Next day, Kennedy met at the White House with NASA's Webb and Dryden, Bureau of the Budget director David Bell, Kennedy aide Ted Sorensen, and science advisor Wiesner. Outside the room, *Life* magazine journalist Hugh Sidey was anxious to learn Kennedy's response to the Gagarin flight. Eventually he was called in to watch as the President staged a little piece of theatre for him. 'What can we do? How can we catch up?' Wiesner, evidently working from the wrong script, suggested a three-month study period to assess the situation, but the President wanted a more urgent response. 'If somebody can just tell me how to catch up. Let's find somebody – anybody. I don't care if it's the janitor over there, if he knows how.' Sidey politely reported a more statesmanlike version:

> *Kennedy turned back to the men around him. He thought for a second. Then he spoke. 'When we know more, I can decide if it's worth it or not. If somebody can just tell me how to catch up ...' Kennedy stopped again a moment and glanced from face to face. Then he said quietly, 'There's nothing more important.'*

As the stage-managed session came to an end, he turned to Sidey and asked: 'Have you got all your answers?'

'Well, yeah, except – what are you going to do?'

Coincidentally, Bob Seamans and Hugh Dryden were in Congress on their second day testifying about NASA's budgets. It should have been a routine number-crunching exercise, but the news about Gagarin transformed the atmosphere. Seamans was assailed from all sides by questions: What were the Russians planning next? He couldn't be sure. Did they want to put a man on the moon by 1967, the 50th anniversary of their Revolution? He didn't know. How soon was America planning to go to the moon ...? Seamans said that in his estimation that goal might well be achievable, although probably not until 1969 or 1970. He was making this stuff up as he went along, trying to keep his answers vague. As he and Dryden stepped outside the hearing room, a barrage of reporters with microphones and cameras was waiting for them. Flashbulbs fired in their faces and once again the questions began. The excitable press folk were all talking at once. 'Can we beat the Russians to the moon?' was the question on their minds. Seamans tried to find a safe path, indicating that it was possible but at the same time it might be difficult.

Then someone asked him: 'Mr Seamans, do you think we ought to put our plans for the moon on a crash basis?'

'It's not for me to make a decision of that sort. That would have to be made by Congress or the President, and basically by the people of the United States.'

Too late, Seamans saw his error. He'd just put Kennedy in the hot seat, possibly forcing him to reply to an issue he didn't actually want to address. At the very least he'd pre-empted the White House. He trudged disconsolately back to NASA headquarters wondering what Jim Webb would think of his clumsy performance. But Webb backed him up, writing a letter to the White House saying that he had done 'an exceptionally fine job' fielding questions in a hostile atmosphere, with 'little or no support' from Democrat politicians who should have known to give him covering fire.

Three days later, the moon was forgotten for a moment when the world learned that Kennedy had suffered another and more serious defeat. A 1,300-strong force of exiled Cubans supported by the CIA landed at the 'Bay of Pigs' in Cuba with the intention of destroying Fidel Castro's communist regime. Kennedy had approved the scheme personally, but Castro's troops learned of the operation well ahead of time and were waiting on the beaches. The raid was a total disaster because the CIA failed to deliver the promised support. Contrary to all expectations, the 'subjugated' population of Cuba showed absolutely no desire to participate in Castro's overthrow. To the CIA's lasting embarrassment, no attempt was made to rescue the invaders.

The Kennedy administration seemed to be faltering in its first hundred days, the traditional honeymoon period during which a new president is supposed to shake things up and make his mark. Kennedy immediately turned to space as a means of reviving his credibility. In a pivotal memo of 20 April, he asked Vice President Lyndon Johnson to prepare a thorough survey of America's rocket effort.

> *Do we have a chance of beating the Soviets by putting a laboratory in space, or by a trip around the moon, or by a rocket to land on the moon, or by a rocket to go to the moon and back with a man? Is*

> *there any other space program that promises dramatic results in which we could win?*

This single-page document can be read as one of the most sensational directives of the 20th century, or else as a hastily dictated panic reaction to a bad week at the White House. As Hugo Young, a journalist from the London *Times*, observed in 1969:

> *Kennedy's response disclosed more than anything the sight of a man obsessed with failure. Gagarin's triumph pitilessly mocked the image of dynamism that he had offered the American people. It had to be avenged almost as much for his sake as for the nation's.*

If Kennedy had all the questions, his Vice President believed he could come up with the answers. Johnson moved at lightning speed to assemble a committee of advisors, including several key business allies: Frank Stanton of CBS, Donald Cook of American Electric Power, and (probably his closest ally of all) George Brown from the Houston-based engineering company Brown and Root. General Bernard Schriever, chief of the Air Force's missile operations, was also involved, keen to argue for an expanded space programme even though his service was unlikely to lead it. Senator Kerr, newly appointed chairman of the Senate Space Committee, was an influential presence throughout. Whatever Johnson and Kerr wanted for space would surely come to pass.

And of course Jim Webb was on hand, bracing himself for Johnson's inevitable onslaught. No matter what happened over the coming days, he 'wasn't going to be stampeded'. A couple of months ago he had been trying to get an indifferent White House interested in space. Now its new-found enthusiasm threatened to hurtle out of control, especially with LBJ leading the cavalry charge. 'He was pressing me, pressing everybody, talking to the newspapers, saying we ought to get going ... He just picked up the phone to everybody that he thought really was the tops. He called Wernher von Braun without even asking me yes or no about it.' Von Braun, of course, was desperate to go to the moon. In a memo to Johnson dated 29 April

1961, he wrote, much to Webb's irritation: 'all other elements of our national space program should be put on the back burner.'

Wiesner knew that Kennedy was about to be railroaded. 'Johnson went around the room saying, "We've got a terribly important decision to make. Shall we put a man on the moon?" And everybody said, "Yes." And he said, "Thank you," and reported to the President that the panel said we should put a man on the moon.' Or to put it another way, in the words of Pulitzer-winning historian Walter McDougall: 'Johnson sent the President a report so loaded with assumptions that a moon landing was the inescapable conclusion.'

And yet Webb, the man who had most to gain from the tremendous expansions of budget and power that must stem from this decision, was the last to climb aboard for the moon. His view was that 'when you decide you're going to do something, and you put the prestige of the United States government behind it, you'd better doggone be able to do it.'

Johnson had no time for caution. 'Why don't you say you'll do it? Why don't you recommend it? Why don't you *advocate* it?'

Webb had already discovered from his new colleagues at NASA that a moon landing was technically possible, and that a great deal of theoretical research had already been conducted, even during the Eisenhower administration. It was also clear that some target in space had to be chosen which the Russians wouldn't be able to match. It seemed they were already capable of lifting substantial payloads into earth orbit, so Kennedy's suggestion of building a space station was risky because the Russians might accomplish that task first. Likewise, a race to fly men around the moon seemed potentially losable so long as Russia continued to have successes with its boosters. Webb had to concede that an actual touchdown on the moon was 'such a big thing that the Russians very probably couldn't do it … The landing could not be fudged. You either did it or you didn't do it. There just wasn't any "if," "and" or "but" about it.' So he did finally let Johnson have his way, but only after insisting repeatedly that 'there's got to be political support over a long period of time – like ten years – and you and the President have to recognize we can't do this kind of thing without that continuing support.'

Dangerous allies

Johnson then urged Webb to get over to the Pentagon straight away and thrash out a deal. No space project on the scale of a mission to the moon could possibly go ahead without the say-so of America's ferocious new Secretary of Defense, Robert McNamara. Barely ten days after first taking up his appointment at NASA, Webb had been immersed in complicated negotiations with him, and the US Air Force, over the development of future rockets and the control of launch facilities in Florida. 'It was a question of whether or not we were really going to be the boss in our own shop, or whether we had to go to the Air Force on everything that we wanted to do. Even if you just wanted to drive a nail in the wall to hang a picture ... We didn't want to spend our time arguing with officials in the Pentagon about what we could or couldn't do.' This most bitter of turf wars had to be resolved, or at least set aside, before NASA could even think about reaching for the moon.

McNamara had gained experience of the military world in the Second World War, plotting the air bombardment of Japan. (He worked with the controversial Air Force general Curtis LeMay.) His business acumen was apparent in his leadership of the Ford Motor Company in the 1950s, from where Kennedy recruited him. He was about to become one of the most controversial figures in American history, the principal strategist of the Vietnam War. But not just yet. For now, McNamara, 'the walking IBM computer' as his critics called him, was concerned mainly with bringing the Pentagon's massive expense sheets under some kind of control. He was unsympathetic to the Air Force's grandiose plans for the conquest of space. The physicist Freeman Dyson, working at that time on an Air Force-sponsored design for a nuclear rocket, quickly sensed the change of mood: 'McNamara was against these Air Force extravaganzas. He wanted the military to concentrate on down-to-earth things, fighting real wars rather than playing around with technical toys. He was always an enemy of the Air Force and more a friend to the Army.'

In light of Kennedy's sudden interest in rockets, McNamara was faced with a dilemma. Fantasies aside, space was an arena with legitimate defence implications, and he had to ensure that proper military concerns were not sacrificed. It was also his duty to make sure that NASA's programmes didn't expand so greatly that they drained away technical talent from vital missile programmes. However, when Webb and Seamans went to see him at his grand office in the Pentagon, they found his technical awareness of space apparently in short supply. According to Seamans, McNamara worried that the Soviets might be capable of a lunar landing at any moment. He suggested that NASA should aim for a planetary trip to Mars instead. Seamans was horrified at the idea. 'Maybe I was cautious, but I didn't think we could possibly do that, even if we'd had the budget. In terms of the technology and our knowledge, we weren't ready then, and even today I'm still not sure we could do it.'

McNamara was no fool, and his dramatic suggestion may have been aimed at diverting NASA with trivial paperwork and theoretical studies for the next decade, instead of building expensive spaceships. Nevertheless, he quickly accepted the benefits of creating real hardware for flying to the moon. Even as he was inflicting strict new efficiencies and cost-cutting requirements on his military contractors, he recognised NASA's space programme as a way of appeasing the aerospace industry without using the Pentagon's money. Technicians assembling moon rockets could just as easily switch to manufacturing additional weapons if the need arose. Until that time came, NASA might as well pay their salaries. McNamara would eventually advise the White House that *without* a lunar programme, many thousands of aerospace workers would have to be laid off.

In that ambiguous spirit – with one eye on the sword, and the other flickering reluctantly towards the ploughshare, so to speak – McNamara and his advisors at the Pentagon wanted NASA's future rocket research to concentrate on easily-storable solid fuels, suitable for missiles on long-term standby in their secret silos, and yet adaptable for space exploration. NASA intended to harness kerosene and liquid hydrogen and liquid oxygen, fearsomely difficult materials to handle, and impossible to store inside a rocket for more than a few hours at a time, but absolutely essential if a rocket was to be sufficiently lightweight and powerful to reach the moon. This turf war over propellants and engine designs had to be settled in NASA's favour. Seamans was impressed by Webb's skilful lawyer's performance as he bluffed, blackmailed and negotiated a document of agreement almost line by line with McNamara's officials, getting rid of every phrase or nuance which he didn't like.

It's notable that Webb wasn't terribly keen to thrash out these terms face to face with McNamara. Throughout his tenure at NASA, Webb would remain more wary of the Secretary of Defense than perhaps anyone else in Washington. McNamara was aggressive, and at the same time, a hard man to read. 'The larger our budget got, the more exacerbating became the desire of the military to have it, and a feeling that somehow space should have been their role ... I think part of McNamara's concern was just a general feeling, "how could anything as big as this be well-run unless I'm running it?" I can't tell you that's precisely what he thought. Nobody knows what he thought.'

For now, at least, an approximate deal was agreed. On 8 May, even as Alan Shepard was being escorted to the White House to be received as a hero, the 'Webb-McNamara Report' was handed to the President: an official peace treaty, so to speak, between NASA and the defence establishment, in which a reasonable distinction was made between liquid-fuelled rockets with exploratory objectives in space, and solid-fuelled missiles which were more obviously of military concern. In principle, NASA was cleared to manufacture giant boosters for the moon without undue interference from the Pentagon. The 'Webb-McNamara' document was a secret one, for it acknowledged certain aggressive facts about the lunar adventure which were not meant to be discussed in public:

> *Our attainments are a major element in the international competition between the Soviet system and our own. The non-*

> *military, non-commercial, non-scientific but 'civilian' projects such as lunar and planetary exploration are, in this sense, part of the battle along the fluid front of the Cold War.*

McNamara and Webb's text also included some controversial thoughts about American business. They wondered if society had perhaps 'over-encouraged the development of entrepreneurs and the development of new enterprises'. McNamara in particular wanted to direct the country's industrial energies more centrally, towards a common benefit, just as the Soviets appeared to have done with theirs. NASA's New Dealer chief shared McNamara's suspicion that *something* was wrong with America as a whole, if it could be so easily beaten into space. On the very day that he heard about Gagarin's flight, he had written a thoughtful note to his predecessor Keith Glennan. 'My own feeling, in this and many other matters facing the country at this time, is that our two major organizational concepts through which the power of the Nation has been developed – the business corporation and the government agency – are going to have to be re-examined and perhaps some new invention made.' Nevertheless, Webb profoundly disagreed with McNamara's idea that there were too many companies going after aerospace work. Competition was a healthy thing, he believed.

An administrator's discount

Kennedy's final decision about going to the moon hinged on NASA literally getting their manned space programme off the ground. On 5 May, US astronaut Alan Shepard was launched aboard von Braun's Redstone booster. His flight wasn't an orbit, merely a ballistic hop of fifteen minutes' duration. Vostok girdled the globe while the Mercury splashed down into the Atlantic just 300 miles from its launch site. No matter. This brief flight was enough to bolster the President's confidence. In a historic speech before Congress on 25 May 1961, he said: 'I believe that this nation should commit itself to achieving the goal, before this decade is out, of landing a man on the moon and returning him safely to the earth. No single space project in this period will be more impressive to mankind, or more important for the long-range exploration of space, and none will be so difficult or expensive to accomplish.' Kennedy's eloquent and famous speech has often been read as a presidential directive. That would be to misinterpret the authority of any president to give such orders. More accurately, it was a question: would Congress agree to his proposal, and if so, would it authorise the funds? The answer to both questions appeared to be 'yes', for the time being at least. Webb found Congress 'in a runaway mood'. He knew it couldn't last, and made his plans accordingly.

All this happened, remember, because Alan Shepard flew just 23 days later than Yuri Gagarin. But Gagarin was first, and the American reaction was inevitable, particularly given the President's driven personality. Logsdon says: 'This wasn't a Soviet success, but an American failure. I don't think it was just a question of Kennedy's responding to public opinion about Gagarin. I think he had his own very personal reaction. He always had a very strong need to be first. He was a very competitive person. He was looking for an opportunity to show leadership and take some kind of bold action.' It looked as if Kennedy had just handed NASA's space cadets their fondest dreams on a plate. They were elated and stunned, and not a little afraid.

Webb now asked for budget assessments. NASA's general counsel Paul Dembling was in Webb's office when the first projections came in. 'He got a figure of $10 billion. He said, "Come on guys, you're doing this on the basis that everything's going to work every time, every place, no matter what you do." So they came back with a figure of $13 billion. So Webb goes up to the Hill with that $13 billion figure.'

Only he didn't. Instead, he calmly informed the politicians that the lunar project would cost upwards of $20 billion. Seamans and his colleagues were appalled. 'Where d'you get *that* figure from?'

'I put an administrator's discount on it', Webb replied blithely.

He knew exactly what he was doing. All the bad news should come at the start of the game, while the politicians were in such a generous mood. If getting to the moon really did cost substantially more than $13 billion – and there was no knowing what kind of unexpected problems or setbacks might have to be paid for somewhere down the line – then at least NASA would look good if it spent no *more* than $20 billion on reaching the moon. Webb had a particular horror of going back to Congress for more money than had been originally requested for a project, and therefore he was looking for a healthy margin at the outset: a margin which Congress undoubtedly would try to claw back in the long years ahead, after its initial excitement had worn off. No doubt the White House's enthusiasms would also shift over time. Once again, Webb warned Johnson: 'I'm ready to go [to the moon], but I must repeat to you that this will require the long term support of you and President Kennedy. There can't be any doubt about that.'

And so, just three months after stepping into NASA headquarters, Webb found himself in charge of the largest, costliest and most ambitious engineering project in human history. As the writer Norman Mailer subsequently observed: 'A giant government bureaucracy had committed itself to a surrealist adventure. Its purpose was clear, but its logic was utterly mysterious.'

Jerome Wiesner might not have put it that way, yet he certainly shared the sentiments. As the senior representative for America's pure research scientists, he believed there was little new knowledge to be gained by flying men to the moon. He salved his conscience by forcing a promise from Kennedy. 'I told him the least he could do was never to refer publicly to the lunar landing as a scientific enterprise, and he never did so.' Many other scientists shared Wiesner's misgivings about the project. Philip Abelson, the editor of the prestigious journal *Science*, was sufficiently angered to say: 'If there is no military value – people admit there isn't – and no scientific value, and no economic return, it will mean we would have put in a lot of engineering talent and research and wound up being the laughing stock of the world.'

Jim Webb had no intention of allowing NASA to become the butt of anyone's jokes. Unfortunately, history had its own warped sense of humour about such matters. On 21 July 1961, astronaut Virgil 'Gus' Grissom flew another sub-orbital curve in a Mercury capsule atop a Redstone missile. His brief trip nearly ended in disaster when the small hatchway on his capsule blew off shortly after he splashed down into the Atlantic. He clambered out of the waterlogged craft without his helmet, and water poured into his spacesuit. He tried to signal an approaching rescue helicopter for help, and was amazed to see it flying over the capsule instead of coming directly to his aid. The helicopter pilots thought Grissom was waving, not drowning. Ignoring him completely, they concentrated on trying to hoist the capsule out of the water before it sank, but it was now so heavy with water that it threatened to pull the helicopter down. Eventually Grissom was rescued, while his capsule sank to the bottom of the Atlantic. The mission was publicised as a success. In truth it was nearly a catastrophe.

Two weeks later, the Russians showcased their celestial superiority yet again. Yuri Gagarin's back-up cosmonaut Gherman Titov had his chance for glory, taking off from the top-secret launch site at Baikonur in Kazakhstan, and flying seventeen orbits aboard the second manned Vostok. His mission lasted an entire day, while Grissom's Mercury had poked its stubby nose above the atmosphere for just a few minutes. America was still some way behind in the space race.

On 20 February 1962, NASA's fortunes improved. The Atlas missile finally demonstrated that it could launch a man into orbit.

John Glenn girdled the world three times and was applauded by US citizens just as Gagarin had been in Russia. At last Webb's agency had arrived as a mature player on the world stage. In this moment of optimism and glamour, he saw his chance to propel his country into the future. 'We were not looking at the cheapest way of making stuff in tents, or building poured concrete buildings. We realized that when we had an engine big enough to leave the air and the earth and to move around in space, we were entering a new and unlimited arena. We were not just constructing the fastest, quickest way to get a few payloads into orbit. We were building permanently.'

Part Two
Bargaining for Power

In the light of the Apollo decision, the Webb-Dryden-Seamans Triad looked at the hardware problems NASA now faced in constructing the new spacecraft. They turned to people with experience in the next most challenging technology: ballistic missiles. The main engineering problem was to source countless components from countless contractors and sub-contractors, and string them all together so that they would work inside one spacecraft. This was a skill for a new kind of manager, the 'systems engineers' who were building Minuteman missiles and nuclear-powered submarine boats for America. So the next question on the table was this: what kind of person should be recruited to spearhead this process?

Webb and Seamans at one point became over-excited about Wernher von Braun as a candidate. After all, his track record so far was excellent. Dryden calmly remarked that this was fine by him if it was what the others wanted, only they could expect his immediate resignation from the Triad if they followed through on the idea. Von Braun was dropped as quickly as he had been proposed. Then there was Abe Silverstein, currently in charge of NASA's manned space flight projects as a whole. He looked forward to the coming Apollo, but disliked von Braun's involvement in it. Unlike so many others in the space community at that time, Silverstein was unwilling to brush aside the V-2 creator's questionable war record. Bob Gilruth, in day-to-day charge of the Mercury project, was also no great fan. He, too, was unlikely to be promoted to the role of Apollo chief. Unfortunately, whoever ran Apollo would still have to be able to deal with von Braun, for he couldn't be sidelined in the coming adventure. He and his greatly expanded rocket team, basking in the glory of launching America's first satellite, and preparing their Redstones to power the first manned Mercury missions into orbit, were now enthusiastically making plans for the gigantic Saturn moon rockets.

Seamans had an idea. Back in his days at RCA, he'd come across a brilliant young electrical engineer, Brainerd Holmes. In 1961 he was near to completing one of RCA's biggest projects, the Ballistic Missile Early Warning System (BMEWS) that guarded against the possibility of surprise nuclear attacks from Russia. The system required radar stations and sensors scattered far afield in England, Greenland and Alaska. Holmes had overseen four big sub-contractors working with a billion-dollar budget and a workforce of around 10,000 people. Countless problems of rough terrain and scattered communications had been resolved so that BMEWS delivered a coherent signal. It was an impressive example of large-scale 'systems management'. Webb consulted with RCA's chief executives, and was told that Holmes was a brilliant man, and they'd only let him go with reluctance. They also warned that Holmes could be a tricky character. But he was a capable young man, no doubt about it. Abe Silverstein accepted a deal whereby he became director for NASA's Lewis Research Center in Ohio, which worked on a wide range of power and propulsion projects. Holmes then accepted the post as supremo of manned space flight. Apollo now 'belonged' to him, and he would make sure everyone knew it.

A fresh-faced 41-year-old with owlish eyes and dark, horn-rimmed glasses, Holmes didn't necessarily have the physical glamour one tends to associate with great leaders. Yet he was a smart, hard-driving man who fought fearlessly to achieve his tasks and drove those around him just as hard as he pushed himself. His track record seemed to demonstrate good management skills; nevertheless, an incident soon after his arrival at NASA set Webb's teeth on edge. When Holmes discovered that he was supposed to report to Seamans as his immediate superior, he said: 'I can't do that. When I was at RCA, Bob was junior to me.' Webb made it clear that the matter wasn't up for discussion. Their terse exchange was a sign of greater conflicts yet to come.

An alliance of rebels

Even as Webb consolidated central management at Washington headquarters, he was thinking about how to keep control over his widely scattered armies in the field. In 1961, the image of NASA in the public imagination was somewhat more unified than the real thing. America's space agency consisted, in truth, of argumentative and independently minded fiefdoms, far apart in geography, history and tradition, each with a different view of priorities in space and on the ground, and each relying on support (or drawing flak) from a different set of senators and members of Congress. Major turf wars were fought between the people who designed rockets in Alabama, the ones who stacked and launched them in Florida, and the teams who built the capsules to fit on top of the rockets, at that time centred in the old NACA heartland of Virginia. In Congress, the elected representatives of Florida fought for funding and influence against the people from Alabama, and so on. Why? Because any given NASA 'field centre' was a source of jobs, and therefore of votes. Centre directors supposedly answered to programme chiefs at NASA's Washington headquarters. If they were ever tempted to circumvent the chain of command by appealing to their relevant state's political representatives, Webb made it clear that he and the Triad would deal with the politicians, and no one else. If everyone inside NASA stayed loyal to the proper internal hierarchy from now on, the benefits would work both ways, helping to protect NASA at every level from an often hostile outside world. 'If there was any undue pressure on a [centre director] to do something he thought wasn't right, I'd say to him, "You go and tell the Congressman or whoever it is putting pressure on you that I told you not to do it, and he should call me."'

It took a while for this message to sink in. When Webb began work at NASA, the most powerful field centre director was, not surprisingly, the great Wernher von Braun. In 1960 his 'Redstone Arsenal' at Huntsville, Alabama, formerly under the Army Ballistic Missile Agency (ABMA)'s control, was renamed the Marshall Space Flight Center and incorporated into NASA. After his successes in launching Explorer 1 and creating the Redstone booster as a reliable carrier for the first Mercury flights, von Braun had won the confidence of many senators and members of Congress. He and his loyal, tight-knit rocket team were ambitious to control America's space effort, even if NASA's nominal chiefs were based in Washington. He was also respected within the Pentagon as a man capable of championing the military benefits of space just as easily as the civilian ones (even though, once installed at Marshall, he actually balked at taking on several Air Force-related development projects which, he felt, detracted from the effort to build the Saturn lunar boosters). A master politician in his own right, he was willing to push against Webb's authority. At the same time – *because* he was a master politician – he always knew when to pull back from direct confrontation.

Another fascinating aspect of von Braun's complicated character, and one that still preoccupies historians of the Nazi era, was his essential respect for chains of command. He would obey orders and show absolute public loyalty to whomsoever his masters were at any given time. Nevertheless, he somehow managed to pursue his own course within this outer appearance of compliance. In a sense, this made him a trickier fellow to deal with than if he had merely rebelled outright at any initiative from NASA headquarters that he didn't like. 'He had the instinct and intuition of an animal', Webb observed of him. 'He could sense danger. You know an animal pricks up his ears and says, "what's going on here, the wind's bringing me a new scent?" Wernher had a remarkable sense of what his audience wanted to hear.'

A couple of examples show von Braun's brilliance at dealing with his supposed superiors. As the Second World War drew to a close, for instance, SS squads were on the lookout for any Germans thinking of surrendering to British or US troops. The rocket team was under guard, nominally for its own protection, but in reality to prevent any of its secrets from being handed over to the Allied forces. Von Braun arranged a meeting with a senior SS officer and persuaded him that if a British or American bomb hit its mark, half his team could be wiped out in a night. He and, more especially, the SS officer, would be in fearful trouble for so carelessly allowing such a valuable asset to be destroyed, von Braun suggested. The best thing would be to split the team into smaller groups and scatter them into different billets, thus reducing the chance of losing the greater proportion of them in a single unlucky bomb strike ... And so, with full SS approval, von Braun scattered his colleagues into the night. The ruse greatly helped their subsequent escape and surrender to US forces.

Another sly move, in later years, saved America's reputation in space. Even as the doomed Vanguard effort was getting under way, von Braun and his ABMA team had been ordered not to launch any Redstone rockets into space. They were to work purely on efforts that contributed to the Redstone's effectiveness as a military missile: the role for which it had been designed in the first place. So the Germans made themselves useful by fitting solid-fuelled upper stages onto the Redstones, tipped with re-entry nose-cones: essentially inert dummies of potential nuclear bombs. It was a perfect opportunity to fly payloads into high sub-orbital trajectories and push pretty hard, at least, against the edges of space. The minor adjustment in the configuration was called 'Jupiter-C', because a 'rocket' of *that* name was not specifically disbarred from sub-orbital flights, while the 'missile' known as 'Redstone' specifically was banned. In 1956, von Braun arranged that the degradation of an entire Jupiter-C rocket could be studied as part of a 'long-term storage test'. The military had to know it could rely on hardware that may have been stockpiled for weeks, months or even years before suddenly being called to fire in times of need. What if the engines corroded or the electronics decayed? How long could these things last? Admittedly, this important test would require that at least one Jupiter-C had to be withdrawn from the nose-cone tests ... von Braun had found a legitimate way of keeping a multi-million dollar rocket on standby for a future space launch that no one had yet authorised. This cunningly preserved booster eventually hurled Explorer 1 into orbit in January 1958, with the Redstone's technology disguised yet again under a new name, 'Juno'.

Webb deeply admired von Braun's dedication to engineering quality and rigorous approach to testing. 'He was very careful to make sure that the rockets would fly. No small accomplishment!' On the other hand: 'He was impatient with paperwork, and he had a tendency to step outside of NASA procedures. He tried to do more at his Marshall center than what his due share at NASA should be.' There was also the matter of public perception, where von Braun consistently outshone Webb. 'You have to remember that he was "Mr Interplanetary Travel". He was known well by a lot of people, whereas people like myself were not known.' Webb was always anxious to keep von Braun in check, yet without stifling his ability to create the Saturn lunar booster.

The next most important NASA facility, in terms of its scale of operations at least, was the launch complex at Cape Canaveral in Florida. Much of this had been inherited from, or built alongside, a sprawl of USAF missile testing stations. It was an uneasy and ill-defined alliance, but that easterly stretch of Floridian coast, centered on the vast, empty swamplands of Merritt Island, afforded both NASA and the Air Force an energy-efficient route towards equatorial trajectories. The earth's circumference at the equator is 24,900 miles. Since the planet spins on its axis once every 24 hours, a rocket at an equatorial launch site benefits from a free ride eastwards at just over 1,000 mph even before its engines are lit; a useful proportion of the 17,500 mph required to reach orbit. In addition, the vast expanse of the Atlantic allows rockets a safe watery graveyard if they veer out of control. For these reasons, NASA urgently needed to consolidate its grip on Merritt Island. Gigantic new hangars, control rooms and launch gantries would be required to stack, fuel and fire the Saturn moon boosters; yet Webb didn't want the Sunshine State's political types to run away with the idea that Florida alone was destined to be America's gateway to space.

Also, as we touched on earlier, there were the people who designed the capsules to ride on top of the rockets: yet another fiefdom with its own preoccupations and desires. A number of far-sighted NACA people had been looking into the possibility of manned space flight long before NASA absorbed their efforts, and none with more idealism and vigour than Bob Gilruth, chief of the old NACA Langley

Center's Space Task Group. Now, under the Keith Glennan regime, he was supervising Mercury on behalf of NASA. Webb had little doubt that Gilruth, a man possessed of great honesty and integrity, should continue in that role. 'Everyone felt they would get a fair shake from Bob. If a problem came to his attention, they'd have a conference and come up with a solution.' Gilruth's tender responsiveness to problems experienced by lower ranks was a model for NASA's ideal: an organisation where no minor functionary or technician was too insignificant to make his or her voice heard. But Gilruth's mild manner and inner core of decency didn't prevent him from holding fierce ambitions. He was concerned that the rocket people at von Braun's Marshall Center might one day encroach on his development of capsules.

Great philosophical differences also existed between Gilruth's people and von Braun's. The Marshall Center was inclined, for instance, towards the fully automatic guidance of rockets and spacecraft. They wanted to *engineer* the progress of a flight from start to finish. Humans were simply payloads to be delivered into space. Since these feeble creatures could not be relied upon to operate in a predictable, quantifiable way, they should be left out of the control loop. This was essentially a 'guided missile' tradition. For Gilruth and his old Langley colleagues, a spacecraft was a piloted vehicle. Theirs was a world in which pilots controlled the success or failure of high-performance machines with verve, skill and split-second instincts. One of the roles of headquarters in Washington would be to help iron out these differences and impose conformity from above, while not at the same time stifling any genuine engineering judgements from people in both camps.

The Goddard Center at Greenbelt, Maryland was another instance of clashing cultures. Its nucleus was the 150-strong Navy Research laboratory team whose Vanguard rocket had so spectacularly failed to catch up with Sputnik in 1957. Now, considerably enlarged, Goddard was primarily responsible for unmanned NASA satellites designed for earth orbit. From 1959 to 1965, Goddard's director Harry Goett oversaw the successful insertion into orbit of 35 satellites, carrying over 100 scientific experiments. In the earliest days of NASA under Keith Glennan, he had also chaired a committee that defined the

Agency's basic ambitions for its future. Goett's performance so far had been brilliantly successful. But this was far from the only quality Webb required of his field centre directors. 'Harry had almost a sixth sense about the contractors you wanted to have for a big project, and yet he wouldn't address the question of how to leave a record that a government agency has to have for doing that kind of work ... You've got to be imbued with the idea that a system of accounting is not a foolish thing. You don't want to thumb your nose at it all the time.'

Goett was a 'czar', the kind of forward-charging supremo who got things done first, and did the paperwork later. He bitterly resented Webb and Seaman's desire to oversee Goddard's affairs. As Webb discovered, Goett 'thought he was so good that he could get away with it – that, by God, nobody could really cause him any serious trouble. We tried, very, very hard ... He said he wanted to draw an absolute line between the people that worked for him, and those that were in headquarters.' Goett battled obsessively with headquarters for the best part of five years, but it was no deal. With Webb's blessing, Seamans eventually asked him to step aside. Goett is remembered affectionately by many people in the space business, but the fact remains that there was no 'right' way or 'wrong' way to run a NASA office in those days. There was only Webb's way. 'He said, "I never thought this would happen to me, because I have such a successful flight record." And my answer to him was, "But you've got to do it right as well as produce success."' Certainly no one would doubt Goddard's track record in successfully launching and controlling a wide range of scientific satellites, including the Hubble Space Telescope (a project in which Webb took an early interest).

As for the old NACA Langley Center in Virginia: great things were on the horizon for the Mercury team in the Space Task Group – just so long as they were prepared to uproot their families and head way down south to the Gulf of Mexico. By 1961, the old NACA facilities were insufficient to encompass development of the Mercury, let alone the complicated lunar spacecraft mandated by Kennedy's epic decision. Webb, Dryden and Seamans recognised the need for a new and specialised 'Manned Spacecraft Center' (MSC) to develop capsules, train astronauts and run missions. This would also counter-balance excessive influence from von Braun's Marshall Center. But was yet more scattering of NASA's command and control strictly necessary? From an outsider's perspective, one obvious option might have been to expand operations in and around the Florida launch facilities, so that booster stacking, capsule assembly, astronaut preparations and the subsequent control of actual missions could be run seamlessly from a single site. It also seemed logical to expand Langley, put in some new buildings and let its Mercury people carry on with what they were already doing, only on a larger scale. Gilruth certainly had no desire to move. He loved Virginia, its climate and people, and above all he would miss Chesapeake Bay, upon which he sailed his own hand-crafted and superbly designed boat as often as he could. Alas, Webb had other plans for him.

The Texas snowball

While Florida and Alabama looked forward to the expansion of NASA centres that they already hosted within their borders, the prospect of an important and brand-new centre triggered a feeding frenzy among politicians from states which, so far, hosted no major space activities. There was pressure from the Kennedy clan to site the new centre on their Massachusetts home turf. Webb's old friend, Democrat Senator William Symington, wanted it for Kansas City in his home state of Missouri, while Senator Hubert Humphrey was angling for Minnesota. Webb had a pretty good idea how this story was going to play out, and a site on the outskirts of Houston, Texas soon became his leading candidate. The Lone Star State's powerful politicians were 'like a snowball rolling down the hill. And you're patting your hands and saying, "Here she comes, boys!" And you think you're doing something about it. But you aren't doing anything about it. The snowball's going to run whether you clap your hands or not.'

Gilruth was startled to learn that Texas was in the running. In late 1961, he and Webb were in an old NACA DC3 transport plane, heading out of Langley to a conference in Washington. Gilruth plucked up the courage to ask: 'Mr Webb, why in the world would you want to make us leave Virginia? Here we've got this beautiful big place at Langley Field, and there's a lot of land, and all that good water around here. I've got my boat. Why do we have to go to Houston, of all places?'

'Bob, what the hell has Harry Byrd ever done for you, or for NASA?'

Gilruth agreed that Virginian Senator Byrd had been no great supporter of space flight, despite having Langley in his back yard.

'That's what I mean!' Webb insisted. 'We've got to get the power. We've got to get the money, or we can't do this program. And the first thing we have to do is, we've got to move to Texas. Texas is a good place for you to operate. It's in the center of the country. It happens also to be the home of the man who controls the money.'

Congressman Albert Thomas was chairman of the particular House Appropriations Subcommittee that determined NASA's budget allocations. Thomas came from Texas, and his district included Houston. Gilruth couldn't argue with that kind of logic.

Webb didn't immediately tell Thomas that NASA was leaning towards Texas because, as so often, he was thinking laterally about his next move. He saw an opportunity to do a favour for the President. At that time, Kennedy was trying (like so many presidents before and since) to steer some new tax legislation through Congress, and Al Thomas was among those blocking the initiative. Offered a bargaining chip by Webb, Kennedy phoned Thomas.

'I need your help on a couple of bills here – three bills.'

'Well now, Mr President, I don't know if I can do that. These are very difficult things.'

'Albert, you know Jim Webb's thinking of putting that new space installation in your district?'

'In that case, Mr President ...'

With this piece of politics, Webb achieved several aims at once. He won yet more favour with Thomas, paving the way for a smooth NASA budget allocation from Congress; he distributed NASA largesse even more widely across the country in his New Dealer style; he gave a clear demonstration, for Thomas's benefit, of his close relationship with the President, while at the same time putting the White House in his debt; and he also ensured that von Braun's people at Marshall could no longer dream of gaining control over manned spacecraft operations. He had put into practice his restless desire 'to accomplish more than one thing with each thing I do'.

While those with a good hand of cards to play dealt themselves in on the NASA game, Harry Byrd from Virginia couldn't be ignored simply because he hadn't fought to bring the new Manned Spacecraft Center to his home state. Not winning it was one thing, but it would be bad form to let the world think he'd *lost* it. Webb took care of the niceties. 'Harry would say, "So, you're moving the Center out of Virginia, and you've come to tell me about it. This is going to hurt me.

How can we handle this?" And I'd say, "Well, we'll handle it the following way, but I have to make the move." He indicated in his own quiet voice that he wouldn't oppose me but he couldn't publicly come out for it.' Webb ensured that Langley's continuing importance as a field centre would not be forgotten, and its staffing levels would remain healthy even after Gilruth and his Space Task Group had deserted it and moved to Texas.

And so the deals were done, and Webb made the announcement for NASA's famous 'mission control' complex. On 19 September 1961, the *Houston Chronicle* ran a delighted banner headline: HOUSTON GETS $60 MILLION SPACE LAB FOR RESEARCH ON MOON SHOT. Most of the front page was devoted to the news, including the fact that around 800 scientists, engineers, technicians and administrative personnel were about to arrive from Langley. The *Chronicle* portrayed Albert Thomas as a man who 'works constantly on behalf of his home community', and said that he deserved most of the credit.

Now, the person who most wanted adoration for putting mission control centre in Houston was Lyndon Johnson. He and Al Thomas were not particularly friendly, perhaps because they both saw themselves as *the* champion of Texan interests. Webb was always sanguine about this kind of puffery. 'We did what we thought was right for the program, and we let the politicians take the credit when and where they wanted to. We never argued with a politician who said, "I got this installation for my state." But we did not accept the judgment of the politicians as to where these installations ought to be.'

At any rate, as Gilruth soon acknowledged, Webb's instincts had been right. 'The people in Virginia weren't as gung-ho for space as they were in Texas. They were thrilled to have us come to Texas.'

We can't help but think of Texas as the home of big cowboy hats, sprawling cattle ranches and ruthless businessmen playing high-stakes games with their oil empires, aided by glamorous shoulder-padded wives and mistresses. This is a myth today, and it was doubly so in the early 1960s. Lyndon Johnson may have acted the big man, but as a child he had experienced the harshest and most debilitating forms of Texan poverty. Not long after Johnson had become Vice President, the British journalist and broadcaster Alistair Cooke received a breathless cablegram from his editors in London:

> WOULD APPRECIATE ARTICLE ON TEXAS AS BACKGROUND JOHNSON. COWBOYS, OIL, MILLIONAIRES, HUGE RANCHES, GENERAL CRASSNESS, BAD MANNERS ETC.

In his subsequent radio broadcast for the BBC, Cooke lost no time in dispelling these myths. 'Cowboys there are, west of the Johnson country, which is central Texas ... but the pasture otherwise is so poor that sheep eat the wild flowers and goats must nibble for the rest ... It is very likely that most of the people here have never seen an oilman. The great oilfields lie to the east and the north, and to Johnson's family and neighbours they are another world. In his early days, in fact, Johnson had to fight the oil lobby on behalf of the poor farmers around him. You could say that the unceasing struggle of the farmers against droughts and floods, their need for some elementary conservation and electricity: these were the hard facts of life that first pushed Johnson into politics.'

As a matter of fact it was an oil company, Humble (the ancestor of today's Exxon brand), that provided the land for NASA's new centre, after considerable wheeling and dealing by Johnson. Webb was happy to let the process play out. Houston's Rice University was gifted a swathe of free land by Humble on the understanding that it would donate at least a thousand acres to NASA. Local businessmen and their supportive politicians were determined to create a benign environment, both for NASA and themselves.

In 1961, the population living in the treeless, salt-grass coastal plain alongside Clear Lake numbered no more than 7,000. By 1964, it had jumped to 30,000. There were 4,900 NASA staffers, along with another 5,000 contractor personnel. Newly appointed Manned Spacecraft Center (MSC) director Bob Gilruth brought the promised 800 people with him from his old Space Task Group at Langley, Virginia. Much of the rest of the human expansion was attributable to Texans eager to take part in the new boom. Today, Clear Lake is home to 175,000 people in nine municipalities. Two million tourists visit every year; many, no doubt, to enjoy the aquatic sporting opportunities of the lake itself, but most to call on the thrumming heart of NASA's mission control.

This great Texan social programme also brought considerable

wealth to certain far-sighted businessmen. Humble Oil may have arranged for the donation of that thousand acres to Rice University, and hence to NASA. What it most emphatically did not do was surrender its controlling interests in large tracts of land within easy reach of the donated lot, amounting to some 30,000 acres. Much of Clear Lake's expansion during the early 1960s was accomplished on that land. Some of it had long been earmarked for oil exploration, but now it was valuable real estate. All those astronauts and mission controllers and ancillary contractors would need houses. John Logsdon, one of America's foremost space historians, is quite blunt when he says: 'It was a real estate scam.'

The Humble–Rice University land deals were subtle, unobtrusive even. They provoked sarcastic comments in the press and among members of Congress, but relatively little in the way of serious flak. The same could not be said of the contract to build the new NASA centre itself.

Twelve miles west of Austin, in the Texas hill country, the Mansfield Dam shoulders the load of 370 billion gallons of water pressing against it from the Colorado River. It was completed in 1941 by the Brown brothers, Herman and George, and their brother-in-law Dan Root. They didn't just build massive concrete walls. They also laid the foundations for the career of Lyndon Johnson. Working inside Congress, LBJ helped get the dam authorised and funded as part of Roosevelt's New Deal reconstruction. George Brown, in turn, made sure that Johnson never lacked for campaign funds. This close and controversial relationship would prosper for the next three decades, taking the Brown and Root company into the space age and beyond. When it was awarded the primary construction of the Manned Spacecraft Center, there were cries of foul play from many quarters. At one point a nervous Johnson called George Brown, expressing concern about unfavourable newspaper coverage. Brown just laughed the whole thing off:

'Hi, George. How are you?'

'Pretty good.'

'I just wanted to check in with you. Did you see that column? They said you had a $500 million one on NASA.'

'Yeah. I just told them I never had talked to you about NASA. I never talked to anybody, never heard of it, but that didn't make any difference. They went ahead and printed it anyway.'

Brown and Root was soon to become part of a consortium which built airfields, bases and hospitals in South Vietnam, and again, it was widely believed that Johnson was facilitating this. Just as a matter of interest: in August 1966, a certain young Republican senator from Illinois was outspokenly critical of this cronyism. 'Why such a huge contract has not been, and is not now being, adequately audited is beyond me. The potential for waste and profiteering is substantial.' In 1967, the General Accounting Office (GAO) identified massive financial improprieties, and disappearances of huge amounts of its war *matériel.* American GIs in Vietnam nicknamed the company Burn and Loot. By the dawn of the 21st century it had become part of the Halliburton group, and Donald Rumsfeld from Illinois was no longer so critical of its construction projects in theatres of war. But that's another story.

No amount of care in the legalities of contracting could protect NASA from charges of favouritism when it came to Texas. In the early 1960s, Jim Webb's support for Texan ambitions was criticised, yet it would be simplistic to imagine that this particular state was entirely undeserving, or that Brown and Root was not a proper company to build the Manned Spacecraft Center at that time. Nevertheless, the story of how Texans got hold of the space age and adjusted it to their advantage does illustrate that NASA was, still is, and probably always will be an organisation whose impact on the ground is far more economically significant than its achievements in space. Even Kennedy became a little concerned about the Texas tilt that seemed to be leaning NASA's expansion its way. He sent for Webb and told him: 'You know, some of my fellows here tell me that you're favoring Lyndon Johnson and Al Thomas on contracts and so forth, and I've got a real problem. We've got to get some work in the bigger cities.' Webb assured him that NASA's selection processes were completely transparent and above-board, and he arranged for mutually agreeable observers to provide some oversight.

On 12 September 1962, Webb, Kennedy and Johnson visited Rice University, where Kennedy made a vigorous speech in support of space: 'We set sail on this new sea because there is new knowledge to be gained and new rights to be won ... Whether [space science] will become a force for good or ill depends on us ...' After he'd made his address, Kennedy couldn't help overhearing his wayward Vice President working the crowd. Webb was appalled to see Johnson 'making quite a speech at the back of the audience, saying that he was going to be sure that everything Texas expected to get out of the space program came their way. Kennedy turned around to me and said, "What's this all about?"'

Texas's good fortune highlighted an all-too-obvious shortfall in NASA's allocations for the heart of the country. In early October 1962, just a few weeks after a much-promoted presidential tour of NASA centres, Governor of Illinois Otto Kerner expressed frustration that the technological wonders of the rocket age had apparently passed his people by. The President's tour had taken him as far into the heartlands as St Louis, Missouri, where the McDonnell company was assembling the Mercury space capsules, yet no further inland than that. 'The production man has always been top dog in this area, and the scientist is a second-rater', Kerner complained. Businesses in Chicago and the Midwest were too preoccupied with tried-and-tested consumer goods instead of innovating for the future. George Beadle, Chancellor of the University of Chicago, backed up Kerner's sentiments with more forthright language. 'The Midwest hasn't responded to new opportunities. We're too busy making toilet seats.' Much of the blame lay with the relevant members of Congress. Illinois, for instance, had just lost its funding bid to host a new solid-state physics lab. Congressmen from Iowa and Missouri had worked to ensure the plan's defeat. Kerner complained bitterly that Illinois' senators 'just won't wheel and deal'. Even McNamara's deputy Ross Gilpatrick had calculated that the Midwest had missed out on some $7 billion's worth of defence business, because of its lack of preparedness. Much has changed today. The Midwest has joined the technological revolution, just like every other square inch of modern America; yet NASA space activities are still thin on the ground in the heartlands, and there are, as yet, no major field centres there. This disproportion continues to affect NASA's fortunes in Congress.

‘Now, *that’s* a politician’

Even as Webb played high-stakes political games in selecting Texas, he tried to ensure at every stage that all NASA’s business was conducted in a proper manner. He was not averse to ‘retro-engineering’ the legalities of arrangements he had already sanctioned, but he always took care that he could tell a straight story if anyone wanted to check the basis of an award. One incident concerning the new Manned Spacecraft Center typifies his skill in reconciling the straightforward needs of his NASA engineers with the more nebulous demands of politics. The story concerns not so much the buildings as the technology they were designed to house.

As the Center began to take shape, its director Bob Gilruth told Webb that a huge new computer system would be needed to calculate orbits and process data from spacecraft. At that time, one company – IBM – dominated the computer market, just as Microsoft does today; and just as with Microsoft, the scale of IBM’s influence sometimes caused discontent among its competitors. Gilruth was adamant that only a trio of IBM mainframe machines could meet NASA’s number-crunching requirements. Webb completely trusted Gilruth’s technical choices, yet he had to think in wider terms before purchasing such high-profile items as the electronic brains for America’s space effort. The problem was simple. If Webb awarded a ‘single source’ contract to IBM just because it was the company best suited to the job, IBM’s rivals would complain most bitterly. They’d file suit in court, perhaps, or send lobbyists to wheedle with the relevant lawmakers in their districts, demanding a chance to make their own bids.

Webb was concerned about the potential reactions of three major manufacturers in particular: Control Data, Sperry and General Electric. He set up personal meetings with the chief executives of all three companies, along with representatives from IBM. 'You're the only four people who can do this. Do you want to bid on this?' General Electric and Sperry dropped out gracefully, immediately recognising that their computers weren't suitable, but Bill Norris of Control Data was confident he could meet the challenge. 'I've got a big powerful machine, I'd like to get this job.' Webb asked him to think things through and come back in three days if he still believed he could fulfil the contract. Norris eventually conceded: 'It breaks my heart to tell you, but I can't.' Webb then placed an order with IBM. He had short-circuited the usual lengthy bidding and selection processes without upsetting anyone, least of all Gilruth, who needed his computers as soon as possible. Seamans was impressed by Webb's finesse. 'Now, *that's* a politician.'

Not so solid rockets

The human space flight programme at NASA holds much of our attention, yet many might argue that the robotic exploration of space has delivered more results, and for less money. Most of America's interplanetary probes are designed and flown by the Jet Propulsion Laboratory (JPL) in Pasadena, California. We are accustomed to its successes (and occasional failures) on and around Mars, and we have witnessed epic voyages to Venus, Jupiter, Saturn and beyond, all managed by JPL, and sometimes involving major contributions from international partners. Jim Webb was called upon to sever NASA's links with JPL at a time when its track record was far less impressive. Fortunately, he resisted.

Founded during the Second World War by the California Institute of Technology (CalTec), JPL pioneered what's known as the 'solid rocket booster', basically a well-made and reliable firework. Solid rockets are ideal for military uses, because they are easy to prepare and launch and can be stored for long periods in silos and hangars, ready for instant use. A young chemist, John Whiteside Parsons, made important breakthroughs in fuel stability, and in the accuracy of the thrusts delivered by what had been regarded as a rather childish technology. CalTec professor Theodor von Karman gave Parsons an opportunity to take his research forward, even though the young rocketeer was more a haphazard inventor than a qualified scientist. The word 'jet' in Jet Propulsion Laboratory disguised what was at that time an untrustworthy field. 'Rocketry' was the stuff of science fiction and schoolboy fantasies. Until, that is, Parsons and his team developed JATO, Jet-Assisted Take-Off, during the war. Small, solid-fuelled rockets attached to bombers allowed them to take off on shorter runways and with heavier loads. JATO's descendants quickly came of age, even as Parsons' wayward career disintegrated in a haze of drugs and pseudo-mystical private pursuits.

It was a JPL team who quickly and efficiently fabricated Explorer 1, America's first satellite, launched atop von Braun's Juno. In addition, JPL's small solid rocket motors worked properly when they were attached between the Juno and the satellite, boosting Explorer 1 into its correct orbit. As military interests in JPL were surrendered to NASA in 1959, it was decided that JPL should remain the province of CalTec, which would run the centre on contract to NASA. Now, this would always make life somewhat awkward, because a contractor is not required to be quite so obedient as NASA headquarters might like. A contractor will supply services and make itself available for quality inspections and review, but it will not necessarily surrender its individual day-to-day management decisions to the NASA hierarchy. CalTec's president, Lee DuBridge, felt that he had greater authority over JPL than NASA. What's more, CalTec charged $2 million a year in licensing fees simply for the privilege of access to JPL's facilities. This accounted for 10 per cent of CalTec's operating budget. Congress balked at the fee, and there were moves to bring JPL directly under federal control and remove it from CalTec's hands.

In Webb's first eighteen months of tenure, five JPL 'ranger' probes to the moon failed in quick succession. The Russians had beamed back pictures of the far side as early as 1959. James van Allen (whose instruments aboard Explorer 1 detected radiation belts around the earth) once commented of JPL: 'They have tremendous esprit de corps. It's almost offensive. It's like the Marines.' Van Allen had identified the core of the problem. JPL was running on militaristic zeal rather than good systems engineering. They were firing off space probes almost like expendable rounds of ammunition to see how well they'd perform. Failures may have been acceptable in secret weapons test programmes, but not for space missions conducted in public view. JPL's early track record was a matter of great concern to Congress.

Webb, of course, believed that any problems were his to cure, not Congress's. He fought off the critics, and in return persuaded JPL's independent-minded director William Pickering to accept some carefully chosen deputy managers who could bridge the gulf between

NASA's way of working and his. 'We felt that if JPL was going to occupy such an important place in our sphere, they ought to do – in addition to the things they wanted to do – those things that NASA needed to round out the programs.' Webb also defended the controversial $2 million annual flat fee accruing to CalTec. In return, DuBridge became a little easier to deal with. The fixes began to work, and JPL gradually assumed responsibility for a succession of unmanned probes to the moon, Venus, Mars, and eventually other more distant worlds. Not all of them succeeded, but those that did scored well-publicised triumphs. Many years later, Pickering wrote to Webb and thanked him for his interventions. 'I thought at one time you were one of the main problems, but I've concluded that we couldn't have done the program without you.'

Pickering might have done well to thank Senator Robert Kerr as well. He helped protect JPL during the 1961–2 budget round, while it was still vulnerable. Seamans was impressed by Kerr's political guile, and by the close relationship that he and Webb obviously shared. Both men seemed to think in similar ways about strategy. When Houston emerged as a leading candidate for the Manned Spacecraft Center, Seamans had a chance to listen in while Webb and Kerr plotted how to exert some additional leverage before making a formal commitment. Even before any names had been brought into the conversation, Kerr knew straight away which man to target and how to play him: 'Okay, you've made the decision. Al Thomas is going to want everything he can get. He's going to want to see that center grow and grow. But be sure you don't give it to him all at once. Make sure that every time you add something in Houston, you exact something that you need for somewhere else. JPL for instance.'

As Kerr knew, in the summer of 1961, with so many political minds turned towards Apollo, JPL and its failing record in unmanned exploration was not a popular subject in Congress. There were moves afoot to cut its funding or even cancel outright the difficult relationship between NASA and CalTec. Seamans was invited to assist Kerr while he spoke on the Senate floor in support of NASA's overall budget request for the coming year. Kerr had finessed events so that the debate would happen on an afternoon when many senators he didn't want to deal with were away on other important business:

namely the All-Stars baseball game. Meanwhile he'd loaded his own bases with supporters. One man who wouldn't be fobbed off that day was William Proxmire from Wisconsin, a state that hadn't as yet shared in the benefits of space. Proxmire proposed an amendment limiting JPL's funds. He began his arguments in a measured tone, but the longer he spoke the wilder he became.

Kerr then spoke in reply, and with little recourse to anything so boring as the facts. At one point he whispered to Seamans, sitting next to him. What did NASA want him to go for? What was the justification for preserving JPL's share of the budget? 'We need more housing', Seamans whispered hurriedly, using the accepted shorthand term for practical working space. Kerr launched into an emotional appeal. As Seamans recalls: 'It was one of the greatest speeches I've ever heard, about the poor scientists and engineers at JPL who hadn't proper accommodations for their families and so on. As he was talking, I tugged at his coat. He leaned down a little bit, and I said, "Laboratories, more space for laboratories."' Kerr segued without breaking stride, and his speech continued as effectively as before.

Eventually it was time for the vote. Compliant senators headed towards Kerr to check which way they were supposed to swing. 'We're against this amendment, aren't we?'

'Oh, yes. Absolutely', Kerr assured them. When a voice vote was called, there were no more than a dozen rebels in Proxmire's camp; and when it came time for the House of Representatives to do its own fine-tuning work on the NASA appropriations bill, there was little complaint from Al Thomas either. JPL's 'housing' was saved.

As Webb sowed his largesse and protection to field centres old and new, so he reaped. The decision to put the Manned Spacecraft Center in Texas coincided broadly with a swift re-ordering of the Washington hierarchy during November 1961. The aim was to withdraw any last vestiges of political power from the field centres and concentrate it in Washington, where Webb could keep a grip on it. From now on, all field centre directors would have to keep in mind their responsibilities to NASA programmes as a whole, rather than just to their individual specialities. Those programmes in turn would be dictated by four main offices in headquarters: one for advanced planning, one for manned space flight, a third for space science, and the fourth dedicated to communications. The variously scattered centre directors would report to their relevant programme bosses in Washington. As far as Apollo was concerned, von Braun and his rocket team in Huntsville were now answerable to an entity called the Office of Manned Space Flight at Washington headquarters, headed by Brainerd Holmes. Likewise, so were Gilruth, in his new Houston complex, and Kurt Debus, chief of the rapidly expanding launch facilities in Florida (and a close colleague of von Braun's). Meanwhile, JPL and the Goddard Center both reported, no matter how reluctantly, to the new Office of Space Science under Homer Newell. All budgets from every programme office would have to be approved by Seamans in his capacity as NASA's day-to-day general manager. Webb and Dryden, meanwhile, would continue to represent NASA's overall interests in Congress and at the White House.

Thinking big

So. Was there more to NASA than a make-work scheme for hungry businesses, a real-estate joyride for Texans or the fulfilling of its role as (to quote the Webb-McNamara memo) 'part of the battle along the fluid front of the Cold War'? Webb certainly believed so. He was among a generation of Democrat politicians and administrators steeped in Roosevelt's progressive New Deal ethos in the 1930s, total world war in the 1940s and nuclear competition in the 1950s. These people had grown accustomed to the idea that entire nations could be run like cohesive machines, or at least rescued from chaos by government interventions. The perfectible society, led by selfless philosopher-kings, is one of the oldest myths of politics, and it has entranced many leaders across much of the 20th century. Already, in these modern times of untrammelled free market economics, we've forgotten the extent to which America once operated a vast 'command economy', and not only to military ends. Just as in Soviet Russia, the overriding belief among the New Dealers was that government influence over the new forces of science and machinery was crucial to national success as a whole.

As budget director for the Truman administration, Webb had tried to clip the wings of the enormous Tennessee Valley Authority, with its 600-mile network of dams and power stations. He had come to regret that stance as a false economy. Now, even as he dealt with the obvious priorities of NASA life – taming the field centres, acquiring the technology, assembling the rockets – he pursued his wider ambition of using space as a platform for social change facilitated by federal funds. No American technological enterprise since the construction of the Panama Canal in the early 1900s, or the building of the immense Hoover Dam in the 1930s, was so grand in its ambitions as Webb's consolidation of NASA in the mid-1960s. Just like its epic predecessors, Apollo was supposed to achieve more than merely its mechanistic function of reaching the moon. At least, that was Webb's vision for the project. Few people working alongside him ever quite grasped the scale of his ideas. He was a 20th-century technocrat in the fullest and perhaps most vainglorious sense. It wasn't lunar dust so much as a transformation of America that he was after. He campaigned throughout the Apollo years for a new brand of 'space age management' in many areas of national life, not just in rocket development. And he wanted this transformation to start with the college kids.

Even after he'd left Washington for a while to join Kerr-McGee, Webb still spent as much time as he could reasonably spare on behalf of government. Despite, or rather because of, the fact that he was a prominent Democrat, he was called on by the Eisenhower administration to serve on several advisory bodies (the Cancer Council, the Institute of Nuclear Studies, the Medical Research Committee and others). Above all, education continued to fascinate him, perhaps not surprisingly, given his background. He served as president of the Frontiers of Science Foundation in Oklahoma, a modest but energetic grant-making body. He also made sure to stop off at universities whenever he happened to be passing nearby on one of his many company trips to Illinois, Florida, Texas, Missouri, Arkansas ... 'I'd take a plane and arrange to meet maybe the chief economist, or the dean of the Liberal Arts school there. I got to be reasonably well-known by the university people. They'd say, "Who is this fellow who drops in when he's passing by on his way to a meeting, who just wants to stop and visit with us for a while?"' Webb had vague but ambitious ideas for a new union between universities, government and the private sector. The ordinary pressures of commercial life prevented him from making much headway with this concept during the 1950s. Now that he was the chief of a giant federal organisation, and possessed of the power to distribute funds, he decided to act.

One of his most cherished projects was known as the Sustaining University Program (SUP). His upbringing and education in 1920s North Carolina had left him with an indelible impression: the Ivy League universities benefited from too much influence over the education system. If candidates had complete freedom of choice as to their pre-doctoral studies, 'they'd all apply to Harvard, MIT and CalTech. Then they'd be turned down, ultimately have to go to Oklahoma or Arkansas or Louisiana, and they'd feel like second-class citizens.' Webb's favoured solution was to create a system of NASA awards, fellowships and grants, so that lesser-known universities could entice the best people onto their campuses. Students wouldn't choose the universities. The universities would choose the students. 'We wanted to turn the whole thing around. We would provide our money for the institution, and they would choose the graduate students in whom they wanted to invest their strength. Another benefit came about. They could then go out and hire a first-class professor, because he knew he had graduate students to work with.'

It was a controversial idea, this paternalistic, top-down management of activities far beyond the business of building machines to reach the moon. Eisenhower had reluctantly signed into law a distant precursor to SUP, the 1958 National Defense Education Act, a response to the Sputnik scare; but as always when it came to Cold War technocratic interventions, he hoped it would prove just a temporary measure. A seven-year limited lifespan was incorporated into the wording of the Act. One of Eisenhower's science advisors, James Conant, had warned him: 'Those now in college will before long be living in the age of intercontinental ballistic missiles. What will be needed then is not more engineers and scientists, but people who will not panic ... not more Einsteins but more Washingtons and Madisons.'

Kennedy, on the other hand, was much more amenable to the linking of space with education. Although reaching the moon ahead of the Russians was his priority, he raised no overt objections to Webb's plan. But then again, he knew few of its details, since Webb buried the budget allocation for SUP within NASA's overall science budget. He wasn't willing to risk debating its merits in Congress. Kennedy's budget director David Bell had queried the SUP. Webb asked him, if he couldn't approve of it, could he at least not object, and allow him legal leeway with a discretionary budget? Bell turned a blind eye, and SUP slipped through the system. This was a profoundly personal agenda. When Congress did eventually discover the SUP spending, there was some concern that Webb was running a programme which no one else knew about. But he got away with (most of) it, and by the mid-1960s NASA's Office of Space Science and Applications was allocating $30 million annually towards university research, with an additional $25 million going direct to the Sustaining University Program. As many as 4,000 new graduates a year benefited from $6,000 each in support from participating universities. These budgets were modest in comparison to the billions spent by NASA on space hardware, yet they were significant by the standards of the day.

A couple of dozen larger individual prizes were also on offer. Eager university deans flocked to Washington's NASA headquarters hoping for one of the 27 'bricks and mortar' grants which were eventually awarded to found new space science laboratories on campuses all around the country. Webb's hope was that academics granted suitable physical resources could create not just theoretical work but actual instruments to be flown in space, thereby competing on an even playing field with the NASA scientists at Goddard, JPL and elsewhere. He was prescient. University labs are now often both the designers and makers of detectors, cameras, scientific experiments, and even entire space probes.

The man in NASA's Washington headquarters supposedly most responsible for day-to-day implementation of the SUP did not share Webb's vision. Homer Newell in the Office of Space Science favoured calling on a wide selection of universities to provide delegates for committees. They would help decide which instruments and experiments were best suited to particular space missions. Newell wanted to draw from the universities' strengths collectively for NASA's benefit, while Webb wanted to shore them up individually, for their own good as much as NASA's. On one occasion Webb was attending the University of Michigan's 150th Anniversary luncheon, and the Dean enthusiastically promised to assemble a committee to help NASA's science effort. He envisaged people from Michigan joining a group from CalTec and Chicago, and – 'Oh no, you don't', Webb interrupted him. 'Not with NASA's money. I want to see what the University of Michigan can do. I want to see you get *your* economists, your historians, your physicists and your electronics people, so that you can say, "This is the best that *our* university can do on these subjects at this time."'

As so often with Webb's machinations, realpolitik was just as significant as idealism. The Sustaining University Program quietened a number of influential academics who might otherwise have complained about NASA's spending on Apollo. Now they, too, had a vested interest in space – just so long as they were prepared to exercise a certain patience towards their patron. Webb couldn't resist micro-managing, especially after Homer Newell's lack of sympathy towards the Sustaining project had become apparent. 'I never realized he had not understood the program that I had laid out for him and put $100 million into ... There was a lack of enthusiasm for carrying it out ... I finally had to take it away from him and do some of it myself.' In 1962, he stipulated that universities applying for NASA grants should try to make contact with a wider range of external organisations and companies in their respective districts. 'I had a very strong feeling that the universities as set up were not going to get enough money from the government, from tuition, or from charitable gifts to do the job of higher education. We had to find a way for them to get, somehow, a proper connection with the profit-making stream of American business.' When Professor Sam Silver of Berkeley requested money for a space laboratory, Webb made it a condition that a pair of economists should be hired to assess feedback between the new lab and its community. It's as if he wanted directly to sense the beneficiaries of his largesse transforming themselves under his fatherly influence.

One might have thought that CalTec and its space laboratory JPL could have served as an exemplar for Webb's university ambitions, especially after he had defended the wayward lab's performance in Congress. CalTec's Lee DuBridge distrusted everything to do with the SUP and vigorously defended his Institute's right to determine its own academic course. Webb, he believed, was a promoter of a dangerous kind of social engineering, and this clash of cultures certainly contributed to his troubles with NASA. Webb discovered that JPL was equally suspicious of his educational ambitions. 'The fact is, there was not a warm, loving atmosphere at JPL for CalTec students who might be interested in space. JPL felt it was easier to do all the work there [in-house] than to take the time and effort to work with graduate students, even from their own institution.' Harvard, not surprisingly, kept its distance too. By contrast, the University of Chicago was much more receptive to Webb's ideas, no doubt in keeping with its Chancellor's desire to haul his city out of the 'toilet seat' business.

One of Webb's special concerns was to break down barriers between specialist university departments in the hope of creating a rich multi-disciplinary atmosphere. He took a great interest in the architecture for the new labs he'd funded. 'If we build enough space in them for the associated disciplines to come and have their offices too, then at least they'd meet in the restroom and pass by in the hall and so forth.' He wanted to encourage universities 'to learn how to deal with big subjects by stretching their total institution, rather than having the disciplines organized below the university level. You see, you've got the physicists here and the astronomers here, and the chemists over *there* – somebody in the government has got to tie it all together.' And that somebody might as well be Jim Webb.

The art of space

And if engineering and the social sciences were supposed to find their perfect union in Webb's vision of the space age, then surely the creative arts couldn't be left out of his thinking. In the autumn of 1962 he initiated the NASA Art Program, which still functions today, albeit in diminished form. The enlistment of reputable artists to make occasional works inspired by space was no great drain on resources. In the age of Apollo, there was room for manoeuvre within NASA's substantial media budget to accommodate the contributions of fine art. The result was a substantial body of work, and one that's still growing today. James Dean began the programme at NASA, while Hereward Cooke of the National Gallery of Art in Washington DC served as the chief curator. Cooke wrote a letter of invitation to a number of artists, using forthright language that Webb must surely have approved:

> *When a major launch takes place, more than two hundred cameras record every split second of the activity. Every nut, bolt and miniaturized electronic device is photographed from every angle ... but, as [French satirical artist] Daumier pointed out a century ago, 'The camera sees everything and understands nothing.' It is the emotional impact, interpretation and hidden significance of these events that lie within the scope of the artist's vision.*

Cooke said that artists would be given access to NASA facilities, and would be subjected to no editorial pressures whatsoever. The only stipulation was that: 'The materials used must be of proven permanence. I do not want faded and flaking pictures in a government archive.'

Paul Calle, famous for his scenes of life in the Old West, took up the challenge, producing superb pencil drawings of astronauts and their capsules. Robert Rauschenberg's semi-abstract silk-screen prints seemed both celebratory and mildly sarcastic at the same time; Bob McCall, a well-known aerospace illustrator, made romantic yet technically accurate pictures to satisfy space hardware buffs, and Lamarr Dodd captured an impressionistic morass of wires, switches, cables and dials, with silver-clad humans embedded into the machinery of their ships. James Wyeth's delicate watercolours showed the forlorn scrublands surrounding NASA's launch centre. Rockets had to keep their distance from the ordinary human realm of towns and streets and family backyards. The drab safety-zone territories around the pads were a sort of endless 'nowhere'. A more inward landscape was explored by Mitchell Jamieson: the psychological drama behind the space age's political rhetoric. It needed only the lightest sweep of an artist's brush to expose the religious and mystical desires implicit in cosmic exploration. Jamieson's image of a geometrically fragmented astronaut in his spacesuit brings to mind the saintly figures in a cathedral's stained-glass window.

Gemini rising

Very quickly after the lunar decision had been made, Gilruth and his manned spacecraft team lobbied headquarters for a bigger and better version of Mercury. Apollo was just paperwork for now, and there was so much yet to learn. How could two spacecraft find each other in the depths of space and link together in a docking? How would astronauts stand up to weeks, rather than hours, in orbit? How could an astronaut venture outside the cabin and 'walk' in airless, weightless space? All these routines needed to be practised. NASA couldn't sit on its hands waiting for Apollo to be built. It needed an interim spaceship in double-quick time. Webb appreciated the value of keeping space in the public eye, and that meant a rapid succession of missions, not just manifestos on paper. Project 'Gemini' was announced in December 1961. McDonnell Douglas set to work on an advanced, highly manoeuvrable two-seater version of its Mercury capsule, while NASA contracted the Air Force to supply 'man-rated' Titan missiles for carrying the load. Other aerospace companies were not too concerned at McDonnell's swift segue into a new contract. There would be plenty of other work on the Apollo.

Webb's reasons for approving Gemini were, as usual, several-fold. Quite apart from the fact that Gilruth's Manned Spacecraft Center had presented a perfectly sound argument, this new ship was his insurance policy. 'I had to have something I could point to if we found Apollo impossible to do. Nobody really knew that you could do it. On the other hand if you finished Mercury, and you did the rendezvous business with Gemini, even if we thought Apollo was impossible we still would have learned a tremendous amount about how to fly and work in space. That is what I was after.' If NASA eventually had to abandon the moon, Webb wanted to get some achievements written onto the score card as soon as possible: long-duration orbital missions, successful dockings and space-walks: the foundations for 'pre-eminence' in orbit at least, if not on the moon.

Gemini attracted the predatory attentions of Robert McNamara. Lurking on the horizon as a problem for both him and NASA was a sinister, black, stubby-winged machine called Dyna-Soar, the 'Dynamic Soaring' space plane. Devised in the 1950s from delta-wing bomber concepts originated by members of von Braun's German rocket team in the Second World War, DynaSoar had already gone through a long and tortured development. It was the US Air Force's most prized manned space project, redesigned now to fit on top of the Titan missile. An expensive full-scale mock-up sat in its hangar waiting for final approval. McNamara couldn't see what good the thing was. It travelled thousands of miles on a long ballistic arc around the earth but could not remain in orbit. Its supposed role as a bomber had become irrelevant in an age of unmanned missiles, while its other potential use as a spy-plane was limited by its relatively short mission profile. Even so, McNamara was properly concerned about America's lack of reliable spy technology in space. The top-secret 'Discoverer' satellites were capable of taking black-and-white photographs of Russian ground facilities. They then ejected a small re-entry capsule to carry the films back down to earth. This was a risky and only partially reliable system. Spy satellites as we know them today, with their much-vaunted ability to zoom in on the tiniest details of a scene from high up in orbit and beam the data down to earth by radio, did not yet exist.

McNamara's favoured solution for the time being was 'Blue Gemini', nicknamed for the blue of an Air Force officer's uniform. He wanted to hijack NASA's new two-man space capsule and use it to service a miniature space station, the 'Manned Orbiting Laboratory' (MOL). A couple of astronauts could squeeze into the MOL for a fortnight or so at a time. The Air Force's powerful Titan rocket, already provided under licence to lift the Gemini, would carry all the hardware. The Pentagon would gain a surveillance capability in orbit, along with a chance to try out new cameras and techniques. The Air Force would be permitted to develop the relatively simple MOL module, so long as they sacrificed the more costly DynaSoar. In time, MOL would also slip off the agenda.

Webb couldn't see how NASA could gain any advantages from Blue Gemini. In fact it would simply get in the way of the capsule's main task: rehearsing the rendezvous and docking procedures which would be essential to the coming moon project. As he made NASA's objections clear, the fragile truce with McNamara disintegrated with alarming speed. Bob Seamans was appalled by the ferocity of the split. 'McNamara had suggested regular meetings, roughly once a month, so that both sides could check in with each other. But this didn't last long. Somewhere around October of 1961, only a few months after Kennedy had decided to go to the moon, there was a terrible argument.'

Webb, Seamans and McNamara were at lunch, along with McNamara's deputy Ross Gilpatrick. Webb was outlining several programmes of mutual interest when McNamara suddenly interrupted him and said: 'Don't you agree that if we're going to have these meetings, we should take up the big issues?'

'Of course. I thought that's what we were doing', Webb replied.

'Then I have some matters I think we should discuss.' And so saying, McNamara flourished a piece of paper on which was written a scathing list of petty grievances against NASA. According to Seamans: 'The problems he listed were hardly astounding. They really were little issues, but he made them sound very damning, as if they proved that NASA was just a dreadful organization. It was the most undiplomatic thing I've ever seen. Jim Webb's face was red with fury. And that was the end of the lunch.' Little wonder that negotiations were conducted largely (although not exclusively) with paperwork or through emissaries from that point onwards. McNamara eventually settled for a token military presence in the Gemini programme. It was no great sacrifice for NASA, since so many of its astronauts, and a good few of its managers on the ground, already had military backgrounds (and the small experimental radar pod carried in the rear of one of the capsules on behalf of the Air Force was also easily accommodated). Webb won the battle for Gemini's freedom after a long and elaborate series of bureaucratic manoeuvres. McNamara remained the one man in Washington he didn't like to deal with in person, in case the two men had a stand-up row. Future NASA administrators would not be so strong in defending NASA's nest against military cuckoos.

As 1962 drew to a close, NASA had undergone an incredible metamorphosis. It had been assigned the world's costliest and most challenging engineering project, and it had begun construction of a giant new space control facility in Texas, while expanding its makeshift launch facilities in Florida into a vast spaceport. Furthermore, it was racing to develop a third manned space project, Gemini, which would come to fruition even before the first Apollo ships left the pad. James Webb had facilitated, and sometimes even imposed, huge decisions with a swiftness that could hardly be imagined today.

The fledgling agency he'd inherited from the Glennan days employed approximately 6,000 personnel on civil service salaries. As Apollo hit its stride, that payroll would expand six-fold, averaging 36,000 people at its mid-1960s peak. And for every NASA staffer, there would be ten personnel in the private sector. The Apollo programme stimulated something in the region of 411,000 jobs around the country. NASA was a federal agency, not an industrial manufacturer. It had no factories, because fabrication wasn't its proper job. Its modest civil service salaries might entice a few skilled managers from industry who wanted to widen their experience, but on the whole, NASA was simply not permitted to pay competitive rates. It was sensible to leave the good people in place, with their perks and company bonuses and all the rest. Webb and his Triad agreed wholeheartedly from the outset that the lunar rockets and spacecraft had to be built primarily by outside industry. Nine-tenths of the Agency's entire $24 billion budget appropriations for Apollo during the coming decade was destined for private businesses. America's powerful aerospace companies competed like wolves for as large a portion of that prey as each could grab.

Winning the prize

Visitors to aerospace museums tend to see the somewhat forlorn heat-seared brown shells of the Apollo command modules. It's now some three decades and more since they flew into space, and they're long past factory-fresh. The world has moved on, and most people's expectations of technology have changed dramatically in the years since men first landed on the moon. Instead of hefty space pods, miniature i-Pods capture our imaginations. The thunderous energies of rocket technology seem crude in comparison to the delicate skeins of the internet or the sub-microscopic subtleties of genetic medicine. People peer through the hatch of an Apollo capsule today and see a dimly-lit interior. The instrument panels are a maze of clunky switches and clockwork dials (no touch-sensitive screens, no plasma displays), and the thousand-fold electrical energies that once powered them are no longer in evidence. Many modern observers are struck by how primitive the capsules seem in comparison to the slick streamlining of modern consumer products. We are reminded repeatedly that a kid's hand-held electronic game is a thousand times more sophisticated than the computers on board those old moonships. And so on ad nauseam.

Perhaps life changes so fast now that our culture has lost the ability to conceive of any world older than last year's. If we want to gain an impression of how futuristic Apollo seemed just three or four decades ago, we need to make a couple of time-travel journeys of the imagination. First, picture the capsule brand-new, completely covered in mirror-smooth silver foil insulation so that it looks like a giant piece of milled, polished metal. (The foil burns off during re-entry, leaving the capsule a drab brown colour. It's best to think of it in its pre-launch finery.) Second, visualise the interior of the spacecraft surgically clean and brightly lit so that the astronauts can see what they're doing. There's not a single scuff-mark, nor the tiniest flaw in the paintwork. The fans and air-conditioning units are humming and the control panel is a shimmer of lights and trembling dials. Apollo is 'alive', and in the mid-1960s it's by far the most complicated machine ever created. Its smooth conical exterior hides some two million separate components fitted together with a watchmaker's precision.

This man-encasing jewel was a product of the North American Aviation plant at Downey in California, eighteen miles south-west of downtown Los Angeles. The twenty or so dirty green buildings which constituted the plant may have looked rather unconvincing on the outside, yet some of them contained 'clean room' facilities where not so much as a single fleck of dandruff could escape the air scrubbers and dust filters. Inside one of these huge workshops, so sterile and bright it hurt the eyes to look at the walls, a dozen capsules were arranged in various states of construction. Technicians in white lint-free coveralls and headcaps crawled and fussed around them like pallid termites attending to their queens. Such extremes of cleanliness were essential. If dust or moisture became trapped inside Apollo's delicate electronics, some vital system might fail. Meanwhile, on the back lot, a tall gantry swung test versions of the capsules over an open water tank, dropping them from a hundred feet to check that they could withstand the expected stresses of impact with the sea during splashdown. Inside the dummy capsules, dummy humans were comprehensively wired up to see how much of a bump their plastic limbs could take. One way and another, the Downey plant looked satisfyingly busy in the mid-1960s.

Founded in 1928, North American Aviation was a small, relatively unknown outfit until James 'Dutch' Kindelberger took control in 1934. The Second World War allowed the company to shine. Its workforce grew to more than 90,000, and by 1945 it had rolled out some 42,600 planes, including the B-25 Mitchell bomber and the famous P-51 Mustang fighter models. At war's end, the company's payroll fell to just 5,000. Kindelberger dabbled in the civilian aeroplane market and dropped an $8 million brick with an unsuccessful machine called the Navion, but by 1950 the company was back on track, manufacturing F-86 Super Saber jet fighters for the

US Air Force. These machines flew thousands of sorties during the Korean War.

In 1954, NASA's precursor, the National Advisory Committee on Aeronautics (NACA) contracted North American to build an experimental rocket-propelled plane called the X-15. Its task was to push all known envelopes of power, speed and altitude in the name of pure scientific and engineering research. It would climb to an adventurous new realm 60 miles up in the sky, where the air became so thin that wings and elevons would lose their purchase and only gas jet thrusters could keep the machine under control, until it plunged back, once again, into a region where the word 'atmosphere' still had some meaning. In all but name, X-15 would be the world's first sub-orbital manned spacecraft.

The problem was, NACA wanted only three of these things. Kindelberger was happier creating huge production lines of fighters, and was wary of creating such a challenging prototype as X-15 if he couldn't then go on to mass-produce it and turn a profit. Anyway, he was a classic stick-and-rudder man with little interest in space or rocketry. He would have passed on X-15 if it hadn't been for a colleague, Harrison 'Stormy' Storms, who saw, perhaps, a farther, higher, faster horizon than his much-loved but ageing boss. Storms knew about space and wanted with a rare hunger to grab a big piece of it. The X-15 project also thrilled him because it was an opportunity to create a wonderful flying machine whose purpose had no immediate connection with warfare and killing. The Mustang and the Saber had been beautifully crafted necessities. This was different. 'Finally we've got a chance to build something that doesn't have any guns on it', he said. Kindelberger was mollified by a $40 million initial payment from NACA and its Air Force backers towards the X-15.

By 1960, even as the X-15 was showing its mettle in the high atmosphere above the Californian Mojave desert, Kindelberger was approaching the end of his time as company chief. To no one's great surprise, Lee Atwood, a trusted lieutenant for more than twenty years, took up most of the responsibility for North American's affairs. Tall, polite and soft-spoken, he didn't seem suited to the rough and tough macho world that was aviation in those days. *Newsweek* magazine described him as:

> *... an introspective, painfully shy engineer with a scholar's tastes. Hanging diffidently at the fringes of a rare cocktail party, driving a modest Chevrolet without a chauffeur, reading voraciously, or skin diving for abalone from a 22-foot boat named* Prospero, *Atwood often seems misplaced in the hard-boiled world of government contractors.*

Soon there would be no place to hide. Some of Atwood's colleagues had grandiose plans for space, and he would be unable or unwilling to rein them in, despite his doubts. There were problems enough with North American's existing space contracts. At a plant in Canoga Park, in the San Fernando Valley north of Los Angeles, a division known as Rocketdyne was grappling with the giant F-1 rocket engine. The Eisenhower administration had approved funding for this thunderous brute even before there existed any rockets large enough for it to propel. Von Braun's tentative designs for the Saturn lunar booster called for a cluster of four, five or perhaps eight F-1s at its base. They were fuelled with ordinary aviation kerosene, but their immense power was proving difficult to contain. When mounted on remote desert test stands and fired up, they tended to explode. This was not good. Rocketdyne's engineers were under intense pressure to fix their 'combustion instability' problems and make the F-1 work. And still Harrison Storms wanted more.

In April 1961, NASA held a conference at von Braun's Marshall Center to announce that a second hydrogen-fuelled stage would be required for the main Saturn stack. Delegates from 30 major companies attended. By the end of the day, only four or five teams were still interested in bidding. The technical challenges were simply staggering. The super-cold temperature of hydrogen in its liquid state plays havoc with the fabric of any structure that tries to contain it. The tiniest variations in the thickness of welded seals in a metal tank are exaggerated to the point of destruction by thermal shock. What's more, hydrogen wasn't the only chemical that the second stage (known as the S-II) had to contain. The combustion process also needed oxygen. To the inexact human imagination, liquid hydrogen stored at 300 degrees below zero seems pretty much 'as cold as' liquid oxygen in an adjacent tank. In terms of the underlying physics, there's a 100-degree difference in temperature between them: as much difference as that between a pan of boiling water and a tray of ice cubes. Yet these fluids in their tanks had to share a common dome-shaped bulkhead, barely an inch thick, in order to make the most efficient use of all the space inside the stage. To cap it all, the structure as a whole had to be ridiculously, impossibly lightweight. According to NASA's brief, the metal skin, pipework and engines had to comprise no more than 7 per cent of the stage's entire weight when fuelled. At the same time, the thin shell of this metal cylinder had to withstand a thrust of six million pounds from below, while carrying yet another 130-ton third stage and lunar ship above it. Undaunted, Harrison Storms was drawn towards the Saturn second stage as a moth is lured to a candle. With Atwood's blessing, North American made a furiously committed bid for the contract to build the stage. And on 11 September, it won.

A couple of months earlier, NASA had also announced its 'request for proposals' for the Apollo spacecraft (the so-called 'command module') to ride on top of the Saturn stack. There was, of course, no question of North American bidding for it now that the S-II second stage prize had been taken. No question, that is, except in the mind of Harrison Storms. As the deadline for Apollo proposals loomed in mid-October, some of the competing bidders were shocked to see Storms and Atwood still heading for NASA briefing conferences. Bob Gilruth had also sent a telegram listing his concerns. Surely North American was busy enough by now?

Atwood was uncertain about bidding for the Apollo. Other companies, such as Martin, Convair and General Electric had already spent a year and several millions of dollars apiece on concept work, positioning themselves for the moon well ahead of NASA's formal requests. He'd considered an offer from the McDonnell company to pool resources, until Storms rejected the plan outright. 'Not interested' was his terse comment. When Atwood demanded that no more than a million dollars of company funds should be committed to the Apollo bid, because the chances of success were so slim, Storms appeared to comply. In fact he was planning to spend a great deal more, by exploiting the inefficiencies of North American's internal accounting system. In those distant days before email, networked desktop PCs and real-time spreadsheets, the various departmental expenses were reported only at the end of each month, after the numbers had been laboriously fed into giant IBM computers using teletypes and punched cards. Once a line item had 'slipped' to the 32nd day of a 31-day cycle, it might not even show up for a couple of months. With a little guile, Storms could buy himself eight or nine weeks before anyone in Atwood's office raised the alarm about his overspending. If he won the Apollo capsule contract, he'd be a hero and no one would complain. If he didn't ... well, that was unthinkable. During a mind-bogglingly intense couple of months in the autumn of 1961, Storms and his troopers drew up a fully detailed bid for Apollo, spending nearly five times their authorised budget in the process. If this gamble were lost, he and many others besides would be out of a job.

On 9 October, NASA's 'Source Evaluation Board' sequestered itself away from interference at the Chamberlain Hotel at Old Point Comfort, near – but not too near – Langley. For the next couple of days, the five leading Apollo bidders were summoned to make presentations. Bill Bergen led a delegation from the Martin company, backed by $3 million's worth of paperwork (a 9,000-page technical report). He was mildly surprised to see Atwood and Storms still in the running at this late stage. 'What are you doing here?' he asked.

Max Faget (the engineer who had first suggested NASA's use of cone-shaped capsules) headed the Board, which was weighted heavily in favour of the old Langley Space Task Group people who best understood how the new capsule should work. There was no favouritism shown to any of the contractors. Far from it. Faget was a brutal interrogator, and Gilruth also made regular appearances, putting contractors under the spotlight with unscripted questions such as: 'What single problem is *the* most difficult task in getting a man to the moon?'

On 24 November, after more than a hundred NASA technical evaluators had pored over the data, the Board made its final report to the Triad at Washington headquarters. 'The Martin Company is considered the outstanding source for the Apollo prime contractor. Martin not only rated first in Technical Approach, a very close second in Technical Qualification, and second in Business Management, but also stood up well under further scrutiny of the Board.' In overall marks out of ten, Martin was awarded 6.9, while North American and General Dynamics shared second place at 6.6, and the two other bidders, McDonnell and General Electric, scored 6.4. It was a comfortable win. Three days after the report was submitted, the good news was announced over the tannoy to delighted workers at Martin's main plant in Baltimore, Maryland.

Storms braced himself for trouble. That $4 million overspend on the bid was surely about to ruin his career. At November's end, while he was away from his desk, his closest colleagues received notice from North American's chief accountants. Would they please report in person the following morning to head office, and be sure to bring all relevant financial records. Storm's ruin merely awaited his return to base. Meantime, he tried to take what comfort he could from a quick trip to Washington to attend an award ceremony at the White House. Test pilot Scott Crossfield was to be honoured for his record-breaking ascents in the X-15. It was good to see North American's rocket plane acknowledged by the President, even if Storm's chances of building any other such machine were now in jeopardy.

Bob Seamans was at the ceremony, and took Storms aside. 'Can you come by my office tomorrow?'

'Well, I'm heading back to LA tonight.'

'Then come by later this afternoon.'

Something was obviously afoot. Storms turned up at NASA headquarters expecting more trouble. Was Seamans somehow gunning for him too? Was NASA worried about the Saturn's upper stage?

'I just wanted to tell you personally, while you were in town', Seamans began. 'You've won Apollo.'

If Storms was rendered speechless for a moment, then his boss Lee Atwood was scarcely less shocked. 'I had no idea we were going to win until it was publicly announced. I was a little surprised, really, having gotten the second stage of the booster some time before, and of course having the rocket engines … I began to realize that, before long, it was going to put quite a strain on the company … I put no personal effort or emphasis or influence into trying to get that command module contract. I thought we had a plate full.'

Doubts were put aside for the moment. North American Aviation was a happy outfit. What workforce won't celebrate when thousands of individual livelihoods are secured for some while into the future? And it wasn't just a question of preserving those old jobs. Over the next two years, Storms' success in winning Apollo triggered an expansion from about 7,500 to 29,500 employees in the company's space division. An often-told rumour had the workers at the Downey plant wearing baseball caps with NA$A emblazoned on them, although this didn't necessarily mean that the company thought it had found the secret to easy days. One shouldn't run away with the idea that Apollo automatically guaranteed large profits. In keeping with many other 1960s aerospace companies, federal work was North American's lifeblood for survival rather than its exceptional bounty. Thanks to Apollo, it was about to become the single largest government supplier. By 1964, defence and space projects accounted for 97 per cent of its $2.2 billion annual turnover. It could expect fees amounting to $4 billion on Apollo alone over the coming decade: a fifth of NASA's *total* budget for reaching the moon. But the lunar spaceships were to be built only in small numbers. There was no prospect of a runaway success in the Apollo capsule market. There was only one buyer on the horizon, and its needs were strictly finite. North American's profits, then, had to come from cost efficiencies during production rather than by clocking up sales of the completed spacecraft. It was a fine juggling act. By the mid-1960s the company was in a surprisingly delicate position, with profit margins of just 2 per cent, or somewhere around $50 million. Any mistakes – $5 million spent on an unsuccessful bid, for instance, or sudden project cutbacks at the behest of Congress – and those profits could easily be compromised.

North American typified the strange relationship that existed between government and private industry in the 1950s and 1960s. It wasn't an expression of pure free market economics, because the government actively intervened in the fortunes of private companies by placing large contracts at the taxpayers' expense, while many of those companies in turn became dependent on government patronage; and the relationship wasn't quite socialist either because the companies remained privately owned, and therefore were permitted to retain their profits. Yet it was without doubt a glossy version of Soviet Russia's 'command economy'. Politicians set the agenda, determined budgets and decided what products they wished to see manufactured and what new technologies they wished to encourage. The thousands of employees in North American's plants and elsewhere were, by any sensible definition, workers for the State. If only 3 per cent of the company's trade was conducted in the private free market, how could it be otherwise? The patchy illusion of capitalist free trade was preserved by the rules of 'competition'. Companies vied for federal contracts by exhibiting cost efficiencies, technical prowess and other fine virtues in their sales presentations. The government then supposedly awarded contracts on the basis of merit and value for money, not favouritism. The illusion had to be preserved that American high-technology companies operated in an open and honest environment. This was best achieved by ensuring fairness in the contract awards. NASA's Source Evaluation Board was a thorough, impartial and highly elaborate mechanism for ensuring that this fairness prevailed. Yet somehow, rumours circulated that North American's victory in the Apollo sweepstakes hadn't been fair. *Newsweek* journalists were by no means alone in commenting: 'There is considerable speculation over just how the company plucked the Apollo plum.'

The exact decision-making process behind the award was not publicly clarified at the time. Webb did not give a full explanation until some five years later, and then only under pressure from an angry Congress. He stressed that it had been an overall Triad decision. He, Dryden and Seamans had overturned the verdict of the Source Evaluation Board, for a number of reasons which all three men thought perfectly valid at the time. For instance, they had listened respectfully to the astronauts, who had expressed considerable dismay when the Martin company scored so highly in the Evaluation Board's report. Webb decided there were significant human factors that the Board had failed to take into account. The men who were about to entrust their lives to Apollo were, as Webb discovered, 'strongly of the view that they would prefer to have a company like North American which had made the X-15 – according to their experience a very high-performance manned aircraft – as against a company who had developed their experience primarily in the unmanned field'. The astronauts' wishes carried considerable weight. Their boss Bob Gilruth was also inclined to trust Storms. As young engineers in the 1940s, the two men had worked together in the old NACA wind tunnels, refining North American's Mustang fighter. The next few years would test their old friendship to the limits. NASA would soon discover the drawbacks of choosing North American to build Apollo.

Part Three
Maintaining Momentum

If Harrison Storms had been pleasantly surprised by the news of the Apollo award to North American, he was nastily shocked when he heard that 'his' ship might not after all be the one to reach the moon. It might fly a quarter of a million miles into space and then be left hanging in orbit a few paltry miles from its target while *someone else's* wretched machine completed the historic journey down to the surface ...

Throughout much of 1962, NASA engaged in a passionate internal argument about the best way to reach the moon with the Saturn booster, or some variation of it. Saturn's development was proceeding apace, and its thunderous F-1 engines were already being tested on the ground, albeit with fitful results. The design of the moon landing ship would determine the eventual size and weight of the booster rocket, and also the number of engines clustered at its base. Different designs called for four, five or even eight of the gigantic nozzles to be fired in harness. The so-called 'mode' decision had to be made quickly, because every aspect of Apollo depended on it. But it wasn't an easy matter.

The long-held dream of simply blasting a rocket to the moon and bringing it back home again was a staple of science fiction, and it couldn't help but infect the imaginations of real-life rocketeers. However, as 1961 drew to a close, the problems of this 'direct ascent' approach had become painfully obvious. The upper stage of the lunar rocket, about the weight and length of a Navy destroyer, would have to land on the moon stern-first without toppling over, and then take off again without the benefit of a launch gantry and ground crew. How else was the crew supposed to get home? It would have to carry all the fuel and equipment for the return voyage, including heat shields for the punishing 25,000-mph re-entry into the earth's atmosphere. Since all this gear would have to go down to the lunar surface and then come all the way up again, the fuel and weight requirements for the whole thing were simply staggering. It was also quite a challenge to work out how the astronauts sitting on the top of this monster were supposed to pilot it safely to a touchdown when they couldn't see the moon through their windows. They'd have to ease the ship down using rear-pointing TV cameras or some such. When over-optimistic sketches emerged showing the rocket landing on its side on skids, that signalled the end of the ball game for direct ascent.

There was much talk, instead, of Earth Orbit Rendezvous (EOR), in which two rockets were launched, and components required for the moonship were docked together in space, before the combined system headed for the moon. This, at least, would save on the crippling weight requirements for each rocket leaving the earth. EOR required precise timings between launches, which no one could guarantee, given the many weeks' launch-pad preparations required for big space boosters. Nevertheless, this vision of an incremental building-block approach to space exploration was pretty much what von Braun had promised in a highly influential series of articles for *Colliers* magazine in the mid-1950s, and it did seem logical. It was also laden with possibilities for future missions, such as space stations and even – one day – trips to Mars. But NASA no longer had time for 'incremental' progressions into space. It had to reach the moon 'before this decade is out'.

During 1961, John Houbolt from Langley and a small group of colleagues within the space industry conceived a lunar landing craft built from ultra-light components. There would be no need for multiple boosters. One rocket could lift all the hardware in a single throw. This idea was known as Lunar Orbit Rendezvous (LOR). The Apollo crew capsule, and all its fuel and other supplies for its return voyage to earth, would remain in orbit above the moon while a separate lander went down to the surface. At the end of the lunar stay, the lander's top stage – a structure so lightweight and fragile you could punch a hole in its skin with a screwdriver – would fire back into orbit and rendezvous with the earth return capsule. Then the lander would be completely discarded, saving even more weight for the homeward trip. The scheme was perfect, except for the extreme hazard of getting the lander to *find* the mother ship: a task requiring far more navigational accuracy than locating the proverbial needle in a haystack. Houbolt's team argued that EOR depended on similarly precise rendezvous procedures anyway, so why not rendezvous in orbit around the moon instead of the earth?

It was a frightening prospect, performing such delicate acrobatics a quarter of a million miles out in space, where astronauts would be beyond rescue if anything went wrong. Von Braun and his team at Marshall preferred EOR, not just because it seemed safer, but because it also guaranteed them a greater technical role in the lunar voyage itself, rather than in just blasting hardware away from the earth. Von Braun also knew that LOR would cancel some of his most cherished dreams. There would be no earth orbiting staging posts on the way to the moon. His lifelong enthusiasm for space stations was founded on a simple instinct. No one would want to leave a perfectly good and very costly station uninhabited, and thus, a long-term space programme would have to be maintained to keep it in service. LOR did not require any such sustained activities in orbit. Quite the reverse. It depended on hardware that functioned for just a few days, hours or even minutes before it was tossed aside and forgotten. However, if the first lunar landing was to be achieved before Kennedy's deadline, then the engineering argument for LOR was unassailable. Von Braun was too honest an engineer to stand in its way for long.

By the summer of 1962, NASA had officially abandoned EOR in favour of LOR. Webb was not particularly alarmed by the change of emphasis, and he made no moves to change any minds within the Agency. As usual, he preferred to trust his people. But Kennedy's science advisor Jerome Wiesner believed LOR was unreasonably hazardous for the astronauts ('They're risking those guys like mad.') and NASA's risk assessments were fundamentally flawed. No amount of technical presentations or mathematical analyses from NASA staffers could persuade him otherwise. He seemed almost irrationally obsessed with hatred towards the scheme. He was assisted by Nick Golovin, a former NASA systems reliability analyst who had kicked up so much fuss about LOR that Webb had fired him. Golovin was more than happy to go over to the enemy. Webb commissioned a blizzard of analysis in response, not wanting NASA to be charged with carelessness in choosing the mode. Nevertheless, he knew that the ultimate decision would be made by NASA and not by outsiders.

North American Aviation also lobbied against LOR. Harrison Storms and his colleagues were disappointed at the idea of another company building the vehicle that actually touched down on the moon. Up until now, it had been assumed that the Apollo capsule itself would be part of the landing package. Wiesner happily listened to some enthusiastic presentations in which the direct ascent concept was reanimated. Meanwhile, the irrepressible Storms announced his intention to bid on the lunar landing module if NASA did decide to build such a machine. Von Braun and Gilruth both thought Storms had his plate full already. They informed North American chief Lee Atwood that no bids from the company would be considered in relation to the Apollo project.

The arguments about LOR became public on 11 September 1962, when Kennedy and his entourage arrived at the Marshall Center for a tour of inspection. Von Braun was proudly discussing the features of a full-scale mock-up of a Saturn booster stage, when Kennedy suddenly turned to him and asked, perhaps not entirely in all innocence: 'I understand you and Jerry disagree about the right way to go to the moon?' If von Braun had any lingering doubts, he wasn't going to express them now. He said, yes, LOR was the way to go. At this point, Wiesner became quite strident, and an argument threatened to break out. A posse of pressmen who thought they'd turned up simply to witness some routine glad-handing between the President and his space experts were suddenly treated to the prospect of a real story: strife in America's rocket programme. But they weren't standing quite close enough to capture the exact nature of the argument. Webb hurried to intervene, and also got caught up in the fracas as he tried to tell Wiesner this was not the time or the place to be arguing against a decision that NASA had already made.

Peter Thorneycroft, a senior British science minister for Harold Macmillan's government, was in the party that day (JFK and the much older Macmillan were close friends). As the presidential plane sped to the next destination (Rice University in Houston, and the seeds of the new Manned Spacecraft Center), a somewhat amused Thorneycroft asked how the LOR debate was likely to resolve itself.

'Jerry's going to lose. It's obvious', Kennedy said.

'Why?'

'Webb's got all the money, and Jerry's only got me.'

Wiesner did have one other influential member of the Kennedy administration on his side. Defense Secretary Robert McNamara also favoured EOR, and for reasons quite similar to the instincts von Braun had expressed throughout his career. The piecemeal assembly of an infrastructure in earth orbit was a surer way of securing permanent strategic dominance in space than the short-term adventures implicit in Apollo. And it's perfectly true that LOR sealed Apollo's demise even before the first Saturn lifted off the pad. Missions that were so self-contained and finite would prove dangerously easy to curtail.

Webb was steadfast in supporting LOR because this flight plan had emerged as NASA's overall technical choice. It was the best available engineering solution to reaching the moon 'before this decade is out', and that was that. He was determined to back up his credo that: 'I couldn't let anybody dictate the decisions that were at the technical level, whether it was the President or the Vice President or the scientists or Mr McNamara.' The wrangling continued for some months, until Webb was moved to visit the White House on 24 October 1962 to settle the matter by announcing his firm intention to let a contract with the Grumman Aviation company of Long Island, New York, for the world's first true ship of deep space: the lunar landing module. Either that, or the White House might like to consider accepting his resignation and putting someone else in charge of NASA. As so often when ridding himself of powerful irritants like Wiesner, he reminded Kennedy: 'Now, look, if you and I stick together, we'll come out all right'; and as usual, Kennedy agreed. At least, this is the impression Webb gives in his various interviews on the final throes of the LOR debate. Kennedy's sign-off may have been somewhat perfunctory that day. Apollo was the last thing on his mind, and even Wiesner couldn't be bothered to put up a fight any more. The Cuban missile crisis was at its height.

'I'm not that interested in space'

Development of the new (and previously unaccounted-for) lunar module threatened to skew NASA's budgets. Costs for the next year, 1963, were now predicted to climb at least $400 million over the limits previously agreed with Congress. An across-the-board 'supplemental' fund would be needed to keep NASA's sprawling multitude of programmes on track. Such was the opinion of Brainerd Holmes, recently appointed as head of the Office of Manned Space Flight in the Washington headquarters, and the senior man responsible for the Apollo project, answerable only to the Triad. Webb's position on the matter was ambivalent. He knew the extra cash could be obtained only if the President threw his weight behind a special request to Congress. Kennedy could claim that the supplemental was justified by urgent national security considerations, but if he did, his reasoning would have to sound good. What if Congress didn't buy it? Webb didn't want to risk embarrassing the White House or, indeed, himself. Anyway, as we've noted, he hated to approach Congress for more money than had already been agreed. He wanted to keep within NASA's annual budget appropriations, so that he could always say: 'We're going to spend what you give us. Make your decisions knowing that this is what we're going to spend. We're not going to be up here asking you for more. We'll come back the next year and ask.'

Just as Webb was calculating what to do about the supplemental, Holmes threw a spanner in the works, claiming that all the extra money should be assigned to Apollo alone. He wanted to cut back on space science programmes within NASA that weren't directly related to putting a man on the moon. Webb flatly disagreed. Although he had recognised the short-term justifications for LOR, he believed that the longer-term explorations of the planets, and other wide-ranging astronomy projects, should not be set aside in the heedless rush to plant an American flag in the lunar soil. NASA's less glamorous missions were just as important for America's future as the conquest of the moon, and perhaps more so, in terms of overall scientific reward. Bob Seamans, trying to be loyal to Holmes and Webb alike, found himself caught in the middle. 'We'd have meetings where Webb would talk about the totality of the program, pre-eminence in space, and so forth. And afterwards Brainerd would come up to me and say, "I don't know what he's talking about. It seems to me our objective is to get to the moon, and I can't put up with all this other crap."'

As Webb found to his dismay, Holmes had even tried to fire the Manned Spacecraft Center's director Bob Gilruth, someone Webb greatly trusted. 'I wouldn't let him do it. That was the beginning of the real rift. He wanted to put his own man in there. A lot of these people want to build a power structure with people loyal to them, rather than to the program.'

Thoroughly frustrated, Holmes went public. In the summer of 1962, an article in *Time* magazine had promoted him as the 'Apollo Czar', fighting against bureaucratic intransigence to keep the lunar landing on track for Kennedy's end-of-decade deadline. Webb was furious, and relations between the two men quickly deteriorated. The crunch came when *Time* ran a second article, 'Space in Earthly Trouble', detailing more of Holmes's complaints. Apparently the difficulties stemmed 'from the clashing personalities and ideas of the [Apollo] project's two top officials ... such are the differences between Webb and Holmes that the whole program is in danger of bogging down.' Webb's staff managed to obtain an advance draft of the story, including a Holmes quote that *Time*'s editors excluded from the final printed version as too strident: 'The major stumbling block of getting to the moon is James E. Webb. He won't fight for our program.'

There might have been some element of caprice in Holmes's mischief-making, yet his arguments did have a certain logic. First, he was in charge of Apollo, which meant he was properly entitled to fight its corner; and second, NASA was required by the 1958 Space Act to carry out national policy in space. Kennedy had made it clear that landing on the moon *was* the national policy, at least for now. From Holmes's point of view, scientific or astronomical investigations that weren't strictly related to Apollo should not be unduly supported within NASA's budgets.

Holmes took it upon himself to warn relevant members of Congress (and especially Al Thomas on the Appropriations Subcommittee) that NASA's formal request for an additional $400 million could be expected any day, with the aim of strengthening Apollo. He also approached the White House with his grievances. An irritated Kennedy, extremely sensitive now to the Cold War ramifications of America's rocket programme, demanded a meeting with him and Webb. On 21 November, freshly incensed by Holmes's latest outbursts in *Time*, Webb arrived at the Oval Office with Dryden and Seamans, his allies in the Triad. Bureau of the Budget director David Bell was in attendance with several colleagues, to keep everyone's minds focused on the dollars. Wiesner was there too, a sceptical onlooker bound to keep most of his opinions to himself after extracting a pledge from Kennedy. ('The least he could do was never to refer publicly to the moon landing as a scientific enterprise.') Unsurprisingly, he didn't have much to contribute that day. Lyndon Johnson was away on business in Texas. He must have been sorry to miss what turned out to be a spectacular meeting. It culminated in a head-to-head argument between Webb and Kennedy about the ultimate purposes of the American space effort.

After a brief flip-chart presentation from Seamans, Kennedy got straight to the point. 'I understand it's a question of whether we need four hundred million dollars more to maintain our present schedule, is that correct?'

'After you made the decision [to go to the moon] in May of last year, we settled on late '67 or early '68 as the landing date', Webb replied cautiously. 'Now this was a target date. We recognized we might have some slippage. We had some financial estimates at that time which have proved to be too small.'

Indeed they were. New requirements had been added to several spacecraft, and a number of contractors had bumped up their fees accordingly. Kennedy wanted to know how the contractor process worked. Webb explained that companies submitted their designs for a spacecraft, based on all the information NASA could provide at the time, and then the best design was selected after an intensive review process. With taxpayers' dollars at stake, competitive pricing was a major factor in a company's bid, but as he pointed out: 'The process of defining [the spacecraft] more accurately usually adds to the cost. And the contractor himself, of course, looks much more carefully when he has been selected. He is not now proposing. Obviously in his proposal he is going to put his best foot forward.'

'So these are not fixed-price contracts?' Kennedy asked.

'No sir, these are not fixed-price contracts.'

'What profit are we allowing them?'

Well, that was a complicated issue. Building space capsules was very different from manufacturing aeroplanes commercially. In the usual way of things, experimental prototypes – jet fighters for the government, say, or airliners for the commercial market – were shaken down in a long series of test flights, until a proven model could be mass-produced for sale to the client. Unexpected challenges inevitably cropped up during the design phase, and the extra costs incurred were absorbed into the unit sales prices of production models. When it came to manned spacecraft, there never was a mass-production run. The skittish experimental versions, numbering no more than a dozen or so, were the finished products and their missions into space had to be considered essentially as glorified test flights. Design changes had to be paid for as and when they emerged. And yes, it also had to be admitted that NASA tended to introduce new and unforeseen requirements almost month by month because they, too, were treading a hitherto unexplored path. As Webb conceded, price hikes weren't always a contractor's fault. 'Where it's clearly a matter that we have added to the contract ourselves to accomplish the mission, it doesn't seem right that we shouldn't allow them their fee.'

To no one's great surprise, the threatened $400 million overspend couldn't be eliminated by curbing NASA's spaceship builders. So, could any programmes be accelerated or made more efficient in other ways? Kennedy asked about progress with the two-man Gemini capsules: essential precursors to Apollo which were under construction at the McDonnell Douglas company. Coincidentally, McDonnell was also near completion on a large order of F-4 Phantom fighter jets. Their construction teams were about to run short of useful tasks. As Webb told the meeting: 'They've got a large number of engineers they'd like to put to work [on Gemini]. So they say, "Why don't we speed up here? We've got resources."'

Webb knew to be wary of such largesse from his contractors. Only so many workers on the factory floor could crowd around the Gemini capsules before getting in each other's way. There was little to be gained by adding more people just for the sake of it. Webb also would have known that McDonnell's offer (backed up with a noisy lobbying effort on Capitol Hill and in the press) was in large part motivated by self-interest. If the Gemini workforce was padded with under-employed Phantom fabricators, it would also become responsible for their salaries, taking the strain off McDonnell's pocket-book and burdening NASA's instead. Webb told Kennedy: 'I sat down with McDonnell and all his people. He said, "I want to spend twenty-five million dollars a month on this program." I said, "Sixteen million a month is all you can spend and we're going to judge you on the cost of this project as well as the technical efficiency." So there you have the difference between twenty-five million he'd like to spend and sixteen million we've allocated.' In a similar vein, the Boeing company had offered to expand its workforce at their gigantic Saturn rocket plant in Michoud, Louisiana, from 4,000 to 7,000 people. Once again, Webb had resisted. 'I would worry about saying to industry, "Just get going, money is no object." Because then you can add not just this four hundred million dollars, but vast sums beyond it … We have been alarmed at these large increases, and have fought tooth and toenail with these people.'

Kennedy conceded the point. NASA's chief was trying to impose firm-minded cost controls, after all. He then pressed Holmes and Seamans for hard dates and mission objectives. When was the first Gemini capsule due for launch? Was that manned, unmanned, what? How many orbits were planned? How soon did they expect a long-duration flight? How many days would astronauts stay up there? A week? Two weeks? His tone was much like that of Premier Kruschev's on the other side of the world, pressing his Russian rocket experts for space triumphs as soon as possible.

Using flip-charts, Holmes explained which milestones NASA could achieve as things stood, and which ones, in his opinion, might be accelerated by the $400 million supplemental. 'It's about a third of a year's difference actually.' Kennedy wanted to know how that four months might specifically affect Apollo. 'I'll give you the next chart, the lunar landing … The effect that this would have is somewhat hard to determine because if everything went the way we anticipated in the design of our Apollo, there'd probably be very little effect on the Apollo …' Holmes, usually so strident, was waffling, but essentially he kept to his line that the supplemental represented, at the very least, 'the only way the schedule could be met'. All being well, the first shot at a manned lunar mission might even 'move to May of '67 by six months from October 1967'.

Jack Kennedy still had everything to play for. That terrible day in Dallas clouds our perspective. We sometimes forget that he had every chance of winning a second term in office and remaining President through to the end of 1968. He knew that if NASA kept on schedule, the first lunar landing might – just might – happen on his watch. It was a prize worth pursuing. After considering how best to approach Congress for the supplemental (and being reminded by Dave Bell that he'd have a tough time justifying it) he turned to Webb and asked: 'Jim, would you say that for four hundred million dollars you'd save six months?'

Webb was blunt. 'I don't think you'd save six months. When it comes to Apollo, I doubt very much it would expedite the landing. This is my personal opinion.' He then explained, a little more tactfully, that Holmes was talking about launch schedules put before industry in order to assess how soon certain objectives could be met, and what resources would be needed. The schedules were discussion tools, not firm commitments.

Seamans backed up his boss. 'I agree with you, Jim, that you can schedule six months earlier, but you have to understand what these dates really are. These are dates for the internal management of the projects. They're by no means dates that you can absolutely guarantee.'

Holmes was losing ground and he knew it. At one point he made a general apology to Kennedy and Webb. He knew the *Time* articles hadn't gone down well. Webb tried, once again, to be gracious: to soften, in front of the President, this damaging impression of senior NASA leaders in conflict. The speculative mission schedule had probably been talked up too much by other senior people at NASA, he explained, and Holmes could be forgiven for taking them too seriously. 'Now, this sort of got cranked up into a feeling that this money was going to be made available, and that a policy decision had already been made to ask for the supplemental. And I think, to a certain extent, magazines like *Time* picked this up in order to make a controversy.'

Kennedy kept his eye on the calendar nevertheless. 'Well, as I at least hear, it wasn't so much that we wanted to speed [the schedule] up as it was how much we were going to slip. You don't like that word, but that's what we're talking about.'

'The reason I don't like the word is that those schedules were never approved by Dryden, Seamans, or me', Webb countered. 'They were not officially scheduled flights in the Agency.' And with that remark, Holmes's authority was undercut once and for all. He wasn't one of the Triad, and it wasn't up to him to confirm any mission dates.

Kennedy then asked Webb a direct question about Apollo. 'Do you think this is the top-priority program of the Agency?'

Webb's answer stunned everyone else in the room. 'No, sir, I do not. I think it is *one* of the top-priority programs.' He started talking about the benefits of science, and some of the other things apart from landing on the moon that rocket technology might achieve. This was definitely not what Kennedy wanted to hear.

'Jim, I think it is the top priority. I think we ought to have that very clear. Some of these other programs can slip six months, or nine months, and nothing strategic is going to happen. But this is important for political reasons ... This is, whether we like it or not, in a sense a race. If we get second to the moon, it's nice, but it's like being second any time. So that if we're second by six months, because we didn't give it priority, then of course that would be very serious. I think we have to take the view that this is the top priority with us.'

'But the environment of space is where you are going to operate the Apollo and where you are going to do the landing.'

'Look, I know all these other things and the satellites and the communications and weather and all – they're all desirable, but they can wait.'

'I'm talking now about the scientific program to understand the space environment within which you've got to fly Apollo and make a landing on the moon', insisted Webb.

'Wait a minute – is that saying that the lunar program to land the man on the moon is the top priority of the Agency? Is it?'

Everyone started talking at once.

Webb then spoke with unbridled passion. 'As we go out and make measurements in space ... the scientific work feeds the technology, and the engineers begin to make better spacecraft. That gives you better instruments and a better chance to go out and learn more. Now, right through all our universities, some of the brilliant, able scientists are recognizing this and beginning to get into this area. You're generating here – on a national basis – an intellectual effort of the highest order of magnitude that I've seen develop in this country in the years I've been fooling around with national policy! The people that are going to furnish the brainwork, the real brainwork, on which the future space power of this nation for twenty-five or a hundred years are going be to made, have got some doubts about it and –'

'Doubts about what? With this program?'

'As to whether the actual landing on the moon is what you call the highest priority.'

'What do they think is the highest priority?'

'They think the highest priority is to understand the environment and the areas of the laws of nature that operate out there.' Webb tried to invoke Wiesner's support for the scientific possibilities of space. 'I think Jerry ought to talk on this rather than me, but the scientists in the nuclear field have penetrated right into the most minute areas of the nucleus and the sub-particles of the nucleus. Now here, out in the universe, you've got the same general kind of a structure, but you can [study] it on a massive universal scale.'

Wiesner was forced to speak, but his lack of support for Webb was startlingly apparent in his patronising tone, despite the NASA chief's overt championing of purely scientific endeavours in space. 'Mr President, I don't think Jim understands some of the scientific problems that are associated with landing on the moon. We don't know a damn thing about the surface ... The scientific programs that find us that information must have the highest priority. But they are associated with the lunar program. The programs that aren't associated with the lunar program can have any priority we please to give 'em.'

Kennedy, exasperated by now, voiced what remains even today a serious question about the value of space flight for the lives of ordinary people. 'Why are we spending seven million dollars on getting fresh water from saltwater, when we're spending seven *billion* dollars to find out about space? Obviously, you wouldn't put it on that priority except for the defense implications.' He reiterated his belief that the only justification for space was that 'the Soviet Union has made this a test of the system. So that's why we're doing it ... The rest of it [the science of space] we can find out about, but there's a lot of things we need to find out about. Cancer and everything else.'

Webb wouldn't budge. 'But you see, when you talk about this, it's very hard to draw a line between –'

'Everything that we do ought to really be tied into getting onto the moon ahead of the Russians', Kennedy stressed.

'Why can't it be tied to pre-eminence in space, which are your own –'

'Because, by God, we keep – we've been telling everybody we're pre-eminent in space for five years and nobody believes it! Because the Russians have the booster and the satellite ... We're ahead scientifically, but – it's like that instrument you got up at Stanford which is costing us a hundred and twenty-five million dollars and everybody tells me that we're the number one in the world. And what is it? I can't think what it is.'

Several voices at once reminded Kennedy about the Stanford Linear Accelerator: only America's most advanced machine, at that time, for probing the deepest subatomic fundamentals of matter and energy.

'I'm sorry. That's wonderful. But nobody knows anything about it!'

By 'nobody', Kennedy meant 'people'. Voters.

'We're not going to settle the four hundred million this morning', he said, winding up the session. 'But I do think we ought to get it clear that [Apollo] is *the* top priority of the Agency, and one of the two things – except for defense – the top priority of the United States government … We ought to be clear, otherwise we shouldn't be spending this kind of money, because I'm not that interested in space. I think it's good. I think we ought to know about it. We're ready to spend reasonable amounts of money. But we're talking about these *fantastic* expenditures which wreck our budget and all these other domestic programs, and the only justification for it, in my opinion, to do it in this time or fashion, is because we hope to beat the Russians and demonstrate that – starting behind, as we did by a couple years – by God, we passed 'em!'

Webb then more or less laid down a threat. 'All right, then let me say this: if I go out and say that [Apollo] is the number-one priority and that everything else must give way to it, I'm going to lose an important element of support for your program and for your administration.'

Kennedy was aghast. 'By who? What people? Who?'

'The brainy people in industry and the universities who are looking at a solid base.'

'I say that the only reason you can justify spending this tremendous – why spend five or six billion dollars a year or whatever we're talking about, when all these other programs, we're starving them to death?'

'Because in Berlin you spent six billion a year adding to your military budget, because the Russians acted the way they did', Webb taunted, using the recent construction of the Berlin Wall as his goad. 'And I have some feeling that you might not have been as successful on Cuba if we hadn't flown John Glenn and demonstrated we had a real overall technical capability here.'

'We agree. That's why we want to put this program – that's the dramatic evidence that we're pre-eminent in space!'

'But we didn't put him on the moon!'

By now, everyone else in the room was in a state of shock. Bell chose this moment to intervene. It didn't sound as if Webb and the President were so far apart in their views, he said. It was just a matter of nuance. Kennedy calmed down, and agreed they probably just needed to clear up the exact wording of an essentially mutual view on space. He asked Webb to write a formal letter expressing NASA's official position on Apollo.

According to Seamans, there was no animosity between the President and NASA's chief that day, despite the raised voices. 'It was just the most dramatic and intense meeting you ever saw. I think Kennedy respected the fact that Jim Webb stood up for what he believed, even if it wasn't what Kennedy wanted to hear just at that moment. Webb represented NASA in concentrated form, if you like. Kennedy could challenge him and see what came back, but I don't think it was ever personal ... I think that after that meeting Kennedy liked Jim's concept that there's more to the space effort than just going to the moon. Anyway I don't believe that he made any major moves afterwards to contradict Webb's thinking.'

And indeed, whatever Kennedy's private thoughts, he stayed true to the gentlemen's agreement he had struck with Webb back in May 1961. Webb could run NASA more or less as he saw fit, unless Kennedy felt threatened by some scandal or disaster in space affairs – in which case, Webb would shoulder all the blame and keep the White House off the hook. That was the ironclad deal, against which Holmes's protests were futile. Over the coming months, NASA's scientific and astronomical programmes continued to enjoy funding, while the inconvenient business of the $400 million supplemental for Apollo was quietly dropped.

And so was Brainerd Holmes. By the summer of 1963, he had become all but ungovernable. According to Webb's recollection: 'He was making efforts to persuade important members of Congress that I was not the proper one to be administrator, and that he was a knowledgeable person and he ought to be put in charge.' At one point, Congress seemed inclined to agree. A closed session of the House Space Subcommittee was convened on 18 June. 'They said, "Well, we think you ought to go, we ought to keep Holmes. You got rid of the wrong man."' Next day's *Washington Post* reported that there had been a three-hour brawl with 'a lot of wrangling and raised voices'.

Webb believed Holmes had gone a little crazy; overcome, perhaps, by the high-profile glamour of running Apollo. 'It appeared to Bob Seamans and myself – Seamans had known him for a long time – that he showed some evidence of instability, of maybe having a nervous breakdown.' But ridding himself of Apollo's wayward 'czar' was not something that could be effected without proper attention to diplomacy. He instructed the NASA press office to announce the termination of the Mercury programme and herald all its successes. Buried deep in this lengthy tract of 12 June 1963 was the news that overall management of the space programme was being 'realigned to permit Mr Brainerd Holmes to return to a position in industry within the period of two years, which was understood to constitute his obligation for government service at the time of his appointment'. Nobody believed for a moment that this ambitious, hard-driving and relatively young man had ever considered limiting his stewardship of Apollo to a mere 24 months, but he toed the official line and resigned with good grace, citing 'personal and financial reasons'. Webb had already asked the RCA company to re-employ him in an influential and suitably well-rewarded executive post so that his removal from NASA would be more palatable to him. 'He chose not to take it. He went to Raytheon and he had a phenomenal success. But that doesn't necessarily mean he would have succeeded in running NASA.'

Many years later, Holmes came to appreciate the qualities of his old NASA boss. 'I never got along very well with Webb. I felt he was a consummate politician with almost no understanding of the technical side of things. However, if I'd been a bit more mature, I would have understood how a politician thinks, and I would have been able to get along with him. I think, overall, he did a very good job.'

'All-up' George

The Triad urgently needed to find a replacement for Holmes. Webb approached the TRW Company, which at that time was substantially involved in missile contracts. TRW put forward George Mueller (pronounced 'Miller') as their best candidate, although company chairman Dave Wright warned that Mueller was likely to prove just as independent-minded as Holmes had been. He had a tendency for fast action which might not always gel with Webb's procedural caution. Mueller was a substantial risk. On the other hand, the temporarily rudderless Apollo definitely needed someone with his kind of drive and determination. Webb decided to try him out. 'If he succeeds, fine. If he doesn't succeed, remove him. This is what the Marine Corps does with its generals. If a general takes too many casualties, they move him out and try another.'

So came to NASA one of its most brilliant and fearless managers. With a somewhat forbidding set to his face, and a tendency to work throughout weekends while expecting his colleagues to do the same, Mueller kept his sentimental side well disguised. From the moment he arrived on 1 September 1963, he tore through every department related to Apollo and challenged all the current assumptions, timetables and ingrained expectations. He quickly concluded that the lunar landing simply wasn't going to be accomplished 'before this decade is out' unless something drastic was done. As he saw it, NASA would be lucky to make 1971.

The problem was that the Saturn V was taking far too long to develop. The German team at Marshall had outlined a careful sequence of incremental tests for the years ahead. The first stage would go up with dummy second and third stages on top of it. If that stage checked out, a first/second stage combination would fly with an inert third stage, and so on. Von Braun believed passionately in testing all components of a rocket as thoroughly as possible before allowing humans to fly aboard it. Mueller argued, instead, for what became known as 'all-up testing'. A fully assembled Saturn should be flown as soon as possible, with every stage stacked and fuelled and ready to go. Why waste valuable time, not to mention expensive pieces of rocket, testing each stage piecemeal? Even the Apollo capsule should be shaken down in an automatic test as soon as possible, albeit with the crew couches empty. Marshall tried to argue that if the Saturn blew up or veered out of control during the test, it might be impossible to determine exactly which stage had failed to light up or which tank had sprung a leak. Mueller said that proper instrumentation on board the rocket should solve that problem. His view prevailed, and once again von Braun showed himself willing to make sacrifices when presented with sufficiently compelling arguments.

Mueller knew that his own arrival at NASA stemmed from previous difficulties with Holmes, and this gave him a certain amount of bargaining power with Webb. He was given a relatively free hand to manage Apollo, so long as he didn't try to encroach, as Holmes had done, on areas beyond his remit. Nevertheless, he made it clear to Homer Newell in the Office of Space Science that Apollo's engineering requirements came first, and science experiments a distinct second. This might have seemed dictatorial, but one only had to compare the possible impact of a scientific payload aboard the Apollo with that of a simple shaver to understand Mueller's desire for control. Early in the programme, Dr Toby Freedman, North American's resident flight surgeon, had been looking at various types of shaver. The batteries in the capsule were too precious to be drained by an electrical model, 'so we worked out a windup mechanical version, using parts of several European models. But then, what do you do about the whiskers you cut off? In space, they'd float all over the capsule, get in people's eyes, gum up the electronic works. So we built a little vacuum cleaner into the razor to suck them away. Well, the whole thing weighed about eight ounces, and that doesn't sound like very much. But it would take an additional 150 pounds of fuel inside the Saturn rocket at the moment of lift-off to account for getting that half-pound of shaver to the moon and back.'

It was little wonder that Mueller felt the need to impose fearsome discipline on the lunar hardware from now on. Webb acknowledged that his new programme manager for Apollo 'didn't cooperate too much with the rest of NASA because he had a big job to do, and he was plowing ahead with that job, and letting the wake of the ship take care of what was left'. He reconciled himself to the need for Mueller's hard-driving, fast-moving approach. Apollo was in a hurry, and he had to trust the people whose very fearlessness qualified them for leadership roles in such a daunting project. He was forced to assume that they would never cross the line between admirable swiftness and reckless speed. Subsequent events would prove how thin that line was.

As for the Apollo spacecraft itself: Mueller recruited Joe Shea and sent him down to the Manned Spacecraft Center at Houston to ride shotgun on completion of the capsule. Shea had learned his trade in the missile business, working for General Motors and then TRW. His task had been to ensure that thousands of components from hundreds of contractors and sub-contractors would fit together inside one machine and work in harmony. Shea was the ultimate 'systems manager'. He began his time at NASA as Apollo's great champion. Before long he would become one of its sacrificial victims.

Keeping order

Just as Holmes had done, the astronauts also discovered in that summer of 1963 that rebellion was useless. It was a matter of some regret to Webb that *Life* magazine had secured semi-exclusive rights to the Mercury Seven's personal stories in return for a $500,000 fee, divided equally among them. While *Life* was an excellent and well-respected publicity platform (if somewhat anodyne), other journals resented their lesser access to juicy stories. Worse still, *Life*'s adulation caused some of the astronauts to believe in their own status as demi-gods, heroes who could do no wrong. Webb appreciated their bravery and lost no opportunity to praise them in public, but he believed most vehemently that they should not place themselves above the thousands of other people who contributed to NASA. Whenever they threatened his authority, he laid down the law. (It's fascinating that in the many astronaut biographies and autobiographies written since the Apollo era, Webb features only as a distant administrator in far-off Washington, influencing their destinies from afar like a capricious Fate.)

Astronauts Shepard and Grissom had opened the Mercury flight rota with brief sub-orbital arcs, while Glenn, Carpenter, Schirra and Cooper had all achieved full orbits. Grissom's capsule had sunk soon after splashdown, although the hapless astronaut was recovered safely, while Carpenter had used up too much of his thruster fuel admiring the view from space, endangering his ability to position his capsule for re-entry. Luck had been on NASA's side, and Carpenter too had come home in one piece. Webb, Dryden and Seamans were now content to call a halt to the programme so that all efforts and funds could be devoted to the follow-on project, the two-man capsule known as Gemini. The Mercury Seven lost no time lobbying for one more flight, for their colleague Donald 'Deke' Slayton had not yet had his chance to go into space. NASA medics had identified a minor heart murmur, yet he and his colleagues were quite sure he was as fit to fly as any pilot needed to be; and even if Slayton was unable to make the cut, the astronauts chafed at the possibility of waiting at least a year, and probably longer, before the next manned flight could be staged using the new Gemini. They pleaded with Webb at least to let Shepard fly a long mission, instead of settling merely for his fifteen-minute hop of May 1961. One more factory-fresh Mercury was on standby, after all. Their request was refused, so they took their complaint to the White House. Next morning, Webb received a wry phone call from Kennedy:

'Well, the boys came by to see me last night.'

'Yes, I know they did. They left my house, told me they were going down to see you, and I told them to tell you everything on their minds.'

'Now, you know who's going to make the decision, don't you?'

'I think I do.'

'You know you're going to make it, don't you?'

'Yes, that's what I thought.'

If Webb wanted to dash the hopes of those epic American heroes in the full glare of press publicity, he had to be seen as the man who took that choice, and not the President. Kennedy was rather more alarmed when newspaper stories began to circulate about the astronauts' financial affairs. They were receiving various benefits from private businesses eager to exploit the glamour of the spacemen. There were free cars, free hotel rooms and more besides. Al Shepard in particular was building himself a canny portfolio of investments and alliances. As soon as the Houston business community got wind of the new mission control complex coming to town, real-estate developers offered the astronauts free homes in order to attract other buyers into adjoining lots. The astronauts, taking risks daily, inconveniencing their wives and families with constant housing relocations, and scraping by on normal pilots' salaries of around $11,000 a year, were more than happy to take up these offers. Some press reports suggested they were encroaching into territories above and beyond their proper entitlements. Once again, Kennedy was eager to keep above all such matters. In an early-morning call to Webb, he made his annoyance clear.

'Say, the *New York Times* is after me, and the astronauts apparently have been offered free houses in Houston, and they're after the White House on this thing. How does the White House get into this?'

'Well now, Mr President, I can't tell you how you got into it, but I can tell you how to get out of it. Just tell them the administrator of NASA is handling this.'

'Well, that's a good idea', said Kennedy.

Webb imposed reasonable limits on the gifts that the astronauts could accept, while preserving their proper rights to exploit their risky profession to achieve at least some degree of personal family security.

The field centre that time forgot

There was one last hurdle to be overcome in the otherwise positive relationship between Kennedy and Webb. Here we encounter the tale of NASA's forgotten field centre. Surprisingly little literature exists, today, on the work of the Electronics Research Center (ERC) in Cambridge, Massachusetts. Proposed during 1963, inaugurated in September 1964, and achieving a peak staffing level of close to a thousand people, the ERC conducted groundbreaking research in communications, navigation, electronic displays, holographic data storage and dozens of other fields until funding was suddenly cancelled outright in 1969. All the newly constructed buildings on the 129-acre plot, just across the street from MIT, were handed over to the Department of Transport. Today, few people remember much about the ERC, for it was something of a failed dream.

We think so easily of rockets as NASA's key technology, yet these were relatively simple and well-known devices. By contrast, the challenges of building computers small enough to fit inside the capsule were not so well understood. Sturdy, vibration-proof circuits had to be devised and tested; the logic of their operations needed to be figured out, and all the celestial mechanics of astronomy, stellar navigation and gravity slingshots reduced to numbers and equations, and then into the bits and bytes, noughts and ones of computer code. Then a 'human interface' between the circuits and the astronauts had to be devised, so that crewmen could 'talk' to these machines, as well as understanding, in turn, what the machines were trying to say. Webb saw these problems as widely relevant to American industry in general, and Cold War superiority in particular. 'The battlefield today is going to be determined more by communications, fire control and electronics. And we are not in very good shape there, in terms of well rounded thinking by people who are concerned with the total government picture.' He was concerned, at one point, that giant communications companies such as AT&T held too much influence over government electronics projects, and that this stranglehold threatened to stifle the innovations which NASA and the rest of the country needed. 'They had such a powerful lock on the Defense Department, and all these different areas of government, that they felt they were controlling the government's use of electronics, and could do it better than anybody else.'

Webb felt otherwise. During 1963, he learned about a failed real-estate deal involving land just across from the MIT campus in Massachusetts. He decided that NASA should found its own electronics research capability there, within easy reach of MIT's brainpower. He also had his eye on Route 128, which snaked its way around much of Greater Boston. Known as the 'Golden Girdle', this otherwise fairly ordinary highway was the umbilicus for Boston's electronics industry. Hundreds of factories were distributed along the Route. Once again, Webb had convinced himself that 'somebody in the government has got to tie it all together'.

Webb had discussed the idea with Kennedy several times during 1962. He was certainly enthusiastic. He did, however, ask that NASA delay making any formal budget request until after JFK's youngest brother Teddy had won his own election to the Senate. The family had held off promoting him until he reached the minimum required age of 30. His campaign slogan was: 'I can do more for Massachusetts.' The Kennedy clan was anxious to avoid charges of nepotism, or of cynically favouring Boston in order to secure Teddy's victory. Webb's chosen tactic for the time being was to bury the $5 million required for the ERC inside NASA's overall $5 billion request for the next fiscal year. The Center would become apparent to Congress only once Teddy was safely installed in the Senate. These were not by any means illegal manoeuvres. NASA was permitted to reallocate modest sums of money internally without having to ask Congress's permission every time; yet this particular shuffle attracted widespread attention. Webb gambled that Congress would eventually see the sense in the ERC: its benefits to Apollo in particular and the national well-being in general. As it turned out, he'd miscalculated. Just for once, his political radar was returning false signals. In January 1963, when Congress finally got wind of the scheme, it rose up and rebelled. The decision to put the ERC in Kennedy's home patch appeared to be 'one of the most controversial research and development decisions of the decade'.

The fact that Webb had tried to sweep the ERC's appropriations under the carpet also didn't help his case. Joe Karth, a Democrat congressman from Minnesota, was particularly outraged. He waylaid Webb and harangued him: 'Look, if you don't tell me that this is just the gutter politics of Teddy Kennedy, I'm going to tell you you're a liar!'

'It isn't, Joe. If you don't believe me I'll take you up to see the President and he'll tell you.'

'I want to go.'

Much to his discomfort, Webb had to deliver on his rash promise. He arranged for Karth to meet the President. Kennedy was readying himself for a speech later that day and was pressed for time. Karth demanded a promise that if he supported putting the ERC in Cambridge, then some other big installation would be built soon afterwards in Minnesota. Webb tried his best to intervene before matters got completely out of hand. 'Joe, the President can't give that assurance. I wouldn't support it if he did.' Kennedy tried to extricate himself from the row and head off to his next engagement, but Karth continued to seethe. 'Joe kept [the President] there unconscionably – I mean, just hammering. You can't just kick a congressman out, but I finally eased him out.'

In March 1963 Webb testified to Congress, giving it some facts of life about the amps, volts and ohms that were just as crucial to Apollo as its rocket motors and liquid fuels. 'Electronics components account for over 40 per cent of the cost of our boosters, over 70 per cent of our spacecraft, and over 90 per cent of the cost of resources going to tracking and data acquisition.' Congress demanded that Webb formally consider other sites for the ERC – Chicago, for instance – before committing to the Boston area. He complied, and special NASA committees led by Hugh Dryden delivered a trio of painstaking reports, all truthful, all valid, and yet somehow all still coming out in favour of that plot of land just across the way from MIT. Although Webb won his victory, the ERC, built on such shaky ground, couldn't outlast his chieftainship of NASA.

In all fairness, he was not alone in favouring the ERC's Massachusetts location. Dryden and Seamans had already agreed with him that a particular man within MIT was essential if Apollo was going to reach the moon.

Once a spacecraft reaches orbit, there's no immediately reliable cue for it to steer by. However, the axis of a freely spinning gyroscope always points in the same direction relative to 'absolute' space. Apollo would be equipped with three gyros angled 90 degrees apart, to serve as a frame of reference. Additional sensors registered the slightest changes in velocity in any given direction. This 'Inertial Navigation System' (INS) could keep tabs on Apollo's position relative to the earth, and especially to the horizon. But the earth itself is constantly moving through space – and the moon too. A fixed external frame of reference was also needed. Before launch, the gyros were to be spun up with their axes precisely aligned with reference to particular 'guide stars'. No mechanical device is perfect, however, and minor misalignments were bound to occur. Apollo also had to be equipped with an optical star telescope so that the astronauts could check for unwanted drift between the gyroscopes and the actual direction of the guide stars. NASA's Triad was united in its view that only one man could design Apollo's intricate navigation system. He was Dr Charles Stark Draper at the MIT's Instrumentation laboratory.

During his wartime years at Sperry, Webb had come to respect Draper as a clever and reliable designer. After the war he had built gyroscope platforms capable of navigating submarines underneath the north polar ice caps, and on round-the-world journeys, without once requiring crewmen to pop their heads out of the conning tower hatches to check where they were. Thousands of miles could be navigated automatically, under the water, to an accuracy of a few feet. Draper was also at the forefront of missile guidance research. NASA's Triad placed a 'single-source' prime contract for Apollo's navigation unit with him and his lab at MIT. Industry complained, but once again, Webb had felt justified in steering NASA along a particular course.

One could almost view Webb as a man who adhered to the rules of government contracting only when it suited him, yet this wouldn't be accurate. He fully respected the rule of law. If he did sometimes load

NASA's technical arguments in favour of choices he'd already made, he did so by exercising his skills as a lawyer, not as a liar. He seldom consciously sanctioned the telling of untruths. On those occasions when it seemed that he might have done – and his greatest test was yet to come – the judgement of history remains divided.

Unexpected overtures

In the autumn of 1963 the White House threw an unexpected curve ball in NASA's direction. Kennedy summoned Webb to tell him about a speech he intended to give to the United Nations on 22 September. His plan was to call on the Russians to cooperate in space instead of competing. This, the President hoped, might defuse some of the international tensions that had so nearly brought catastrophe during the Cuban missile crisis. It might also allow space funding to be trimmed at a time when many Americans were beginning to question Apollo's staggering expense. Kennedy's about-turn was triggered, in part, by comments from Sir Bernard Lovell, the internationally renowned radio-telescope astronomer from Britain. In a letter to Hugh Dryden, Lovell reported his discussions with senior Russian scientists in Moscow. Apparently Russia had no plans to go to the moon. Instead, it intended to focus on manned space stations in earth orbit. There was no way of verifying this, and Russia's scientists may simply have been spouting officially approved falsehoods. It didn't help when a Soviet scientist disingenuously informed Dryden that the Academy could speak only for itself and its scientists. It couldn't be responsible if some bureau elsewhere in the vast Soviet system still had lunar ambitions. But the simplistic message relayed by the American press was that Apollo might be made redundant even before the first mission was launched.

Kennedy sensed that a diplomatic manoeuvre could disarm media criticism and widen his options when it came to space. He summoned Webb and told him he would propose a joint Russian–American lunar mission before a full session of the United Nations. 'Are you in sufficient control to prevent my being undercut in NASA if I do that?' he asked bluntly.

If Webb was taken aback he tried not to show it. 'Sir, I have sufficient control, and I will see that you are not undercut.' He immediately phoned all NASA's field centre directors to prevent them making any more speeches such as the one that Gilruth had made a few days before, suggesting that Russian–American cooperation was unrealistic. Their orders, now, were to 'make no comment of any kind on this matter'.

Of course cooperation wasn't going to happen, at least not yet. None of the American pieces of hardware could be mated with its Russian counterpart. The fuel and gas pressures were different, the electrical systems weren't compatible, and NASA's digital computers were entirely unlike the Russians' simpler electrical 'sequence controllers'. Webb knew this perfectly well. The United Nations speech was a political gesture that might, perhaps, result in some shared scientific probe or astronomy project. It certainly couldn't deliver on the President's romantic challenge at the UN to 'explore whether the scientists or astronauts of both countries – indeed, of all the world – cannot work together in the conquest of space, sending some day in this decade to the moon not the representatives of a single nation, but the representatives of all our countries'.

The Russians' response to the speech was one of almost complete silence. They would not wish to share the secrets of their best rockets, or expose the comparative lack of sophistication inside their capsules. This came as no great surprise to Webb. It was the reaction at home that caught him off-guard. 'I didn't think it would be taken by the people in Congress as a reason for withdrawing their support. I was a little surprised at the violence of the reaction down there. It didn't seem to me this was necessary. It seemed to me that a president ought to be able to put that kind of thing forward in a speech at the UN for discussion on a worldwide basis without the people who were supporting the program saying that he was backing away from it.' Webb began to realise that the President might not have been entirely honest with him. 'Maybe there were other factors that I didn't fully understand. I took Kennedy's word on the basis of full faith that he was [making the speech] for the reasons he gave me. There were indications that people around him wanted this to be a slight withdrawing of support [for Apollo].'

It's a widely held myth that money was no object for Apollo throughout the 1960s. As early as the end of 1963, as NASA's budget requirement for the coming year was debated in Congress, Webb's grand plans came under fire. It wasn't just the man himself but federal largesse towards space in general which seemed, to some, alarmingly un-American. In November, William Proxmire, a consistent thorn in Webb's side, was moved to suggest that 'the space program is probably the most centralized government spending program in the United States. It concentrates in the hands of a single agency full authority over an important sector of the economy ... [This] could well be described as corporate socialism.' A few days earlier, Arkansas Democrat William Fulbright, founder of the renowned Fulbright Scholarship programme, spoke about a 'dangerous imbalance between our efforts in armaments and space on the one hand, and employment and education on the other'. Congress awarded $600 million less than the (admittedly staggering) $5.7 billion that NASA had requested: not the kind of result Webb was used to. He had to call on help from Al Thomas, still safely installed on the relevant House Appropriations Subcommittee. The two men embarked on a feverish behind-the-scenes campaign to limit the damage. Additional cuts amounting to $900 million had also been tabled by Congressional critics, and if these had been approved, Apollo's schedule for the moon would have been severely jeopardised. Just as Webb had predicted in 1961 when first accepting the NASA chieftainship, grand promises made by America's political establishment in one year were not necessarily sustained into the next. His 'administrator's discount' on the total price of Apollo had been prescient indeed.

Well, if that's what they wanted on the Hill, Webb would make sure the lawmakers took responsibility for their own budget choices. As always, he refused to go back later, cap in hand, begging for 'supplemental' funding for some programme or other, if Congress had already turned him down. 'They'd hand you this "hush, darling." They'd say, "Gee, this is about three or four hundred million dollars less than you need, take it and get going and come back in three months and we'll give you another one." I felt that if we were going before committees of the House and Senate every year, and [submitting to] the budget process every year, then we couldn't do any more. So I made it a rule, "We will not ask for a supplemental, Mr Congressman. When you decide you want to make a cut, that's what you're going to live with, because we're going to make the cut."'

And where was Kerr? Dead of a sudden heart attack on the first day of 1963. Webb sorely felt his absence. Undoubtedly, he would have fought Webb's corner if he'd still been around, but his Senate Space Committee was now in the less generous hands of Clinton Anderson from New Mexico. While Anderson was a fellow Democrat, and supported space in general, he certainly did not favour Webb. In fact, he wanted to cut him down to size. As the NASA budget debates came to a head, he and other Webb critics on both sides of the partisan divide were cocking their ears, listening for faint noises off-stage. Skeletons that had remained buried while Kerr was alive were beginning to rattle now that he was dead.

Left exposed

Webb's final meeting with Kennedy occurred in early November 1963. As the President geared up for re-election, Webb wanted to discuss how space might play on the campaign trail. Bruised by the United Nations incident, and with Blue Gemini much on his mind, he forewarned Kennedy that McNamara was still proving difficult. 'I have to tell you, I think that the Secretary of Defense will not want to support the [space] program as having substantial military value. So you're going into a campaign with me saying it has very important technological benefits for the military, and the Secretary of Defense being unwilling to say it.' Webb's message, albeit coded, was that he wanted presidential authority to block McNamara's interference.

'Well, is there anything personal between you?' Kennedy asked, knowing full well that there was. 'Don't let it get personal.' He assured Webb that he'd handle McNamara, and on that note of mutual understanding – Kennedy insisted he was still Apollo's leading supporter – the two men parted for the last time.

NASA could survive McNamara's obscure manoeuvrings, and Webb was strong enough to handle Congress without Kerr's help, but it would prove much tougher to survive the loss of his single greatest protector. On 22 November 1963, President Kennedy was assassinated in Dallas. The man who'd once said 'I'm not that interested in space' had spoken truthfully, yet Webb had spent three very productive years using him as a shield. Anyone who attacked him was attacking the President. He was under direct orders from the President. The President had told him to get to the moon. The President hadn't told Webb *not* to do any of the other things he was trying to do, and if anyone had a question about Webb's authority to do what he was doing, why, go ask the President and see what *he'd* say. Now the

President was gone. Undoubtedly this was a gut-wrenching loss, yet it also delivered a grim kind of opportunity. From now on, if Webb's enemies in Congress tried to hinder Apollo, he could accuse them of betraying the sacred legacy of a dead president.

The President was dead. Long live the President. A shocked Lyndon Johnson was sworn into office that same day. Of course he could be counted on as a NASA supporter, although as Webb knew only too well: 'Once he became President, he had a different set of problems than before. He wasn't quite as free to press for those areas he had a particular interest in.' A man campaigning for the White House could afford to champion individual causes as a sign of his vigour and determination to win. A man *in* the White House had to balance countless priorities at once, because this was the principal task of a presidency.

Johnson's first years as President were marked by a boldness of vision unprecedented since the Roosevelt era. Substantive civil rights legislation was steered through Congress. The Great Society and the War on Poverty expressed his passionate commitment to improving the lives of poor people. He was an unreformed New Dealer. Even as he promised to carry on the work left uncompleted by Kennedy, it was Roosevelt who truly continued to inspire him. Alas, the money wouldn't stretch to Johnson's noble visions. Signs of change became apparent in the first year of his regime. As he tried to push $11 billion's worth of tax cuts through Congress in a bid to strengthen the economy (an initiative inherited from the Kennedy administration), his most obstructive opponent was Harry Byrd, the very same senator from Virginia whose lack of zeal for space had helped persuade NASA that Texas might be a better home for its Manned Spacecraft Center. On one occasion, Johnson called a meeting in the Cabinet Room of all the heads of major federal departments, including NASA. He said to each, one by one, that he wanted a 5 per cent cut across the board, so that the tax cuts could be accommodated. Webb was out of town on business and Bob Seamans was NASA's point man that day. When Johnson asked him if he could comply with the cuts, he said, carefully: 'Of course we'll do everything we can to make the reductions you've asked for, but I'm sure you wouldn't want to jeopardize the lives of the astronauts in so doing.'

Bill McKee, head of the Federal Aviation Administration, caught up with Seamans as they were leaving the White House. 'You ought to know that the President just said to me, "Will you take that young man Seamans out behind the barn and give him the facts of life?"'

Johnson subsequently met with Webb to discuss his problem. 'I've just got to get some kind of a tax bill through, and Harry Byrd won't support it unless I guarantee to hold NASA expenditures under $5 billion, and I want you to do that.'

'Alright, sir. If that's what you want me to do, that's what I'll do.'

Webb had no choice but to agree to Johnson's demand. He knew that other areas of government apart from NASA were also suffering cuts. The wind was changing, and that was that. As Johnson struggled with Vietnam, so his passion for space inevitably became diluted. He remained in favour of it, and instinctively committed to Apollo, but the need to occupy the 'high ground' of space in perpetuity against some terrifying Russian cosmic threat no longer seemed so pressing. The 'Domino Theory' seemed more urgently alarming to the White House strategists than the prospect of a 'Red Moon'. The American public, meanwhile, was increasingly divided about the virtues of fighting Communism by proxy, whether in space or on the ground. Protecting the Free World against Russia in the 1950s and early 60s had been widely accepted as necessary. The merits of a distant war in Vietnam were less obvious, and most especially to a post-Second World War generation now coming of age.

Webb took care to lend a sympathetic ear throughout Johnson's presidency. They were both Southerners and committed New Dealers, after all. They held many dreams in common. 'He and I had a more intimate relationship than I ever had with Kennedy. In many ways, he was sort of a lonely man. Every once in a while he'd want to get someone that he knew and trusted, and call you up and just talk on the phone ... He loved to call you up and just talk. He also was a great man to try you out, to put a great deal of pressure on you just to see how you reacted to the pressure. He got a little fun out of that, I think.'

NASA's fortunes were subtly on the wane, yet the momentum that Webb had already built up would sustain the Agency: not just for now, but through one of the worst crises in its history. Meanwhile, he was anxious not to allow his empire to calcify into a complacent, self-serving bureaucracy. He believed that a modicum of instability would keep people on their toes. 'My attitude was, if you have 200 project managers who were adequate, you should still remove ten per cent of them every year to keep the pressure on them to do better … If a project manager feels he has a permanent job, then you've lost the whole impetus of what you're trying to do with the project. He's got to cold-bloodedly finish his project and take his chances on the next assignment.' Webb made continual adjustments to middle management contract departments, industrial relations experts and so forth. He might have done better to keep a closer eye on the senior figures of Apollo and check what they were doing.

According to Seamans: 'Webb really wanted to have the management of NASA as his final, great undertaking. People would say, "How do you manage something?" They'd say, "You do it the way Jim Webb did." He was always fretting and agonizing and tweaking and trying to change the management. And at times, he had someone working for him – namely me – that wasn't always too enthusiastic about that. I thought we had a tremendous job to do, and that we could get the job done pretty much with the organization we had.'

One of Seamans' problems after 1965 was that he could no longer count on Hugh Dryden as a buffer between him and Webb. Dryden had withstood the pain of cancer for many months without complaint, but by the autumn of that year he was no longer well enough to work. At the beginning of December he died. Seamans became NASA's Deputy Administrator, and he moved into Dryden's old office immediately alongside Webb's. He soon realised how much pressure Dryden had absorbed and deflected on his behalf during the lifespan of the Triad. Now he felt exposed, and less able to counter Webb's intractable will. NASA had lost its wise elder counsel, and Seamans was surrounded by men whose ambitions were in danger of blurring their judgement. Above him was Webb, ever restless, ceaselessly demanding; below him were Mueller and Shea, forging ahead with Apollo on their own terms and at a dangerously fast pace.

The seductions of success

The two-seat capsules that flew between Mercury and Apollo in the mid-1960s don't get much of the glory in space history today, but Gemini was probably NASA's sportiest craft, with its huge gull-wing doors for space-walking or escape by ejection, and its complex array of thrusters. It was also the first space vehicle to carry an on-board computer (with fifteen kilobytes of memory) to calculate its rocket firings. It enabled Gemini to change orbits and accurately locate other target vehicles for docking: the most essential of all space manoeuvres.

After two successful unmanned shots, Gemini 3 took off on 23 March 1965, with Virgil 'Gus' Grissom and John Young at the controls for a five-hour orbital test. Grissom knew the Gemini better than anyone, and he always felt that the ship was a calculated risk. In a sombre moment shortly before launch, he told his wife: 'If there's a serious accident in the space program, it'll probably be Gemini, and it'll probably be me.' Only a few days earlier, he had alarmed Webb by being a little too honest at a pre-flight press conference. 'If we die, we want people to accept it. We're in a risky business.' Grissom's luck held for now, and Gemini returned safely to earth.

On 3 June, Jim McDivitt and Ed White spent four days aboard Gemini 4 with a blown computer. McDivitt had to plot his re-entry calculations with help from the ground. Meanwhile, the world's attention was focused on the first NASA space-walk, conducted by White. He floated outside the capsule for half an hour, connected only by a snaking umbilical cord. 'This is great! I don't want to come back inside!' he said.

In August (the launch rate was extraordinary), Gemini 5 proved that humans could survive long enough in space for a round trip to the moon. Gordon Cooper and Pete Conrad got the job, sealed for eight days inside the capsule. Power failures dogged the mission, and Conrad worried about the retro rockets in the rear. Neither astronauts, nor their capsules, had ever been asked to fly for this long before. What if the retro rockets succumbed to the cold? What if they didn't work on day eight to bring them home? In private, Conrad decided that if his ship got stranded in orbit he'd slit his wrists rather than suffocate ... Luckily he had no need to worry.

These Gemini missions, we have to remember, were Webb's main exposure to actual manned flights. The man most responsible for laying the groundwork for Apollo witnessed only one successful earth-orbital test of that craft before stepping down from his post. Gemini was also President Johnson's opportunity to show off, for he, too, would be out of office before Apollo achieved its principal goal. Now was the time for the big Texan to capitalise on 'his' NASA. Even as Conrad and Cooper shivered in their capsule, he was making big plans for their return to earth. Bob Seamans discovered this when he was abruptly hauled into the White House for a dose of the Johnson treatment:

'Sit down! Seamans, you guys over there at NASA, you're pretty good with your science, and you're pretty good up there on the Hill, but as far as I'm concerned, Seamans, you're a great big zero. You know why you're a great big zero, Seamans?'

With Seamans' zero-ness safely established, the two men got down to the nitty-gritty. It turned out that Webb had blocked Johnson's desire to send some of the Gemini astronauts on a global goodwill tour. Seamans was ordered to get hold of Webb, who was on a mountain hiking tour in Carolina with his family. Both of them were to report to Johnson's Texas ranch the coming weekend. They arrived, exhausted and wary, only to find that Johnson had set up one of his little ambushes. His public affairs officer Bill Moyers was holding a text that the President had already concocted without consulting NASA.

'I've got the press release about the astronauts' goodwill tour here, but I want your comments on it, Jim. We'll let Bill read it.'

Moyers began. 'After the most successful space flight of all time –'

Webb cut in. 'It wasn't that successful, Mr President. We had some trouble with the fuel cell.'

'Alright, Jim', Johnson said. 'After the least successful space flight of all time ...' Webb squirmed as Johnson gave a broad wink to Seamans.

And the upshot was, Conrad and Cooper went on their goodwill tour.

An unmanned Atlas-Agena target vehicle was hauled aloft at the end of October for the next Gemini to link up with, only to blow up six minutes into its flight. Wally Schirra and Tom Stafford were on an adjacent pad in Gemini 6 waiting for lift-off, and had to abandon their mission. Gemini 7, carrying Frank Borman and Jim Lovell, departed on 4 December for a gruelling fourteen-day medical experiments session. The thrusters glitched, the electrical power systems glitched again, and it looked as if two weeks was more than the little ship and its uncomfortable crew could deliver. The last three days of the mission were almost unbearable, and Borman was keen to come home, but with plenty of support from ground control he kept his head.

With the first Agena target vehicle blown to smithereens, Gemini 6 was rescheduled to rendezvous with Gemini 7. Even if the two ships weren't actually equipped to dock, NASA could still pull off a coup. The Russians had shown that they could launch pairs of capsules into orbit in such a way as to pass briefly within hailing distance of each other, but so far they hadn't managed to manoeuvre any of them into close contact. They lacked both the computing power and the complex array of thrusters required to manipulate their orbits and nudge their ships into a more intimate union. 'No way can the Russians rendezvous two spacecraft', said Schirra. 'All they can do is, they see the pretty girl on the pavement from their car window before the flow of traffic whizzes them on. In Gemini we can cut across the traffic and say hello. Now *that's* what you call a rendezvous.'

The reworked mission got off to a hair-raising start when Gemini 6's Titan rocket suddenly shut down on pad. According to the rules, Stafford and Schirra should have ejected, just in case the rocket blew up, but they kept their cool and nothing happened. Their bravery prevented the Gemini cockpit from being burned out by the explosive charges under the ejection seats. All the hardware survived, and the ship got off the ground three days later on 15 December. Six hours into the mission, Stafford and Schirra steered to within less than five feet of Gemini 7, so close that they could see Lovell and Borman grinning at them through the other ship's windows. NASA was completely in its element, or so it seemed.

Back down to earth

It wasn't just the astronauts who were on an accelerated learning curve. Seamans believed that the best way to do his job was to visit the field centres and other facilities as often as possible, show his face, and listen to problems from people on the ground. He wasn't the kind of man who liked to run everything from his office. Indeed, he was struck by how seldom Webb strayed from Washington. Nevertheless, there were occasions when it was best to stay close to the office, and especially when things went wrong somewhere within NASA's great sprawl on the ground – or in space. In all his time at NASA, Webb attended only one blast-off from Florida. 'I saw the Gus Grissom launch [aboard the second manned Mercury capsule], and then I went and picked him up on Grand Turk Island after they fished him out of the water.' This messy and near-fatal splashdown convinced him that 'if you have any trouble anywhere it is more important for me to be at the center of communications, and available to the President, than it is to be down there. And besides Dr Dryden liked to go to the launches, Dr Seamans liked to go to them, and so I just let them go because they wanted to go.'

Seamans soon discovered the wisdom behind this. Neil Armstrong and Dave Scott blasted off aboard Gemini 8 on 16 March 1966, to dock with an unmanned Agena target vehicle. It was the first test of NASA's ability to lock two craft securely together: an essential skill for reaching the moon. Astronaut Bill Anders was just about to start his shift in the mission control room as Capsule Communicator, the one person on the ground who actually talks to the astronauts in orbit and relays instructions. (The so-called 'CapComs' are always astronauts.) As he strolled into mission control to start his shift, the docking was under way, and colleague Jim Lovell handed over the microphone, saying: 'It's all pretty boring so far.' Moments later, Armstrong's voice came on the line. 'We've got serious problems. We're tumbling end over end.' Armstrong immediately backed his capsule out of the Agena, only to find that the tumbling became worse. A thruster was jammed in the 'on' position and Gemini 8 began to spin wildly out of control. At 90 revolutions per minute, Armstrong and Scott could barely keep conscious …

At that moment, Seamans was arriving at a black-tie dinner in Washington, an annual event celebrating the work of one of America's earliest rocket pioneers, Robert H. Goddard. 'When I went to the dinner, everything was just fine. The docking took place on schedule. It was going very well. I didn't have a radio in the car to listen, and nobody called me on the telephone in the car. So when I drew up to the hotel, a gang of NASA people descended on me and swept me into a room and briefed me on what was going on. I was told we were in deep trouble.'

Lyndon Johnson's Vice President Hubert Humphrey was due to make the keynote address. Seamans was greeted by 'very long faces'. Everyone in the room by now had heard that something was going wrong up in orbit. He got on the phone to NASA headquarters for an update, and then had the unpleasant task of announcing to everybody at the dinner that 'we had this situation that could have a devastating result. People didn't believe it at first, and then of course the whole tenor of the dinner was sort of electrified.' Humphrey began his speech, and after what seemed like an interminable wait, a phone call came in with the news that the Gemini crew had stabilised their spin. Seamans whispered the news to Humphrey, who announced it to the assembled guests.

After the dinner, Seamans found, unsurprisingly, that 'Mr Webb was not very happy with my performance, feeling that we'd taken sort of a gamble there; that we were announcing publicly what was going on – or I was – when we didn't really have the facts well in hand.'

The occasional dramas in the Gemini programme simply served to emphasise how great its successes were, as men began to familiarise themselves with the challenges of outer space. Meanwhile, development of Apollo appeared to be progressing well under the strong leadership of George Mueller. Of course there were dozens of technical problems, budgetary complications and schedule scares; yet the general impression, both within NASA and Congress, was that such problems were manageable. Properly trained astronauts could *manage* their balky thrusters, and properly marshalled resources on the ground could *manage* the building of Apollo. The dangers of space were calculable and avoidable, or at least, survivable, by virtue of NASA's correct management of risks. NASA was a trusted organisation in the mid-1960s, and Webb's reputation in Congress still carried weight, even against his critics. He began to talk, more and more, of space age management as a general tool for solving some of America's wider problems on the ground.

Seamans felt uneasy as Webb continued to experiment with his broader philosophy. 'Jim would ask one of the Apollo construction managers, "What do you think your responsibilities are?" And he'd get this reply, "Well, to build the capsule and make it fly." And he'd say, "Isn't it much more than that? Can't you see further than that? Don't you also want to try and help with the satellites, or some other aspect of what we're trying to do at NASA?" And a lot of people struggled to get that vision. I shared it to a large extent. I tried to rationalize what impact our program as a whole would have in a wide range of areas. But I didn't share it to quite the extent that Jim believed in it. Maybe he sometimes went a little far.'

Webb decided to re-invent the management structures at Washington headquarters just at a time when the existing modus operandi appeared to be producing satisfactory results. An old friend from his State Department days had suggested to him that reorganising NASA in mid-flow was like 'trying to perform an appendectomy on a laborer carrying a heavy trunk up three flights of stairs and getting it done before he reaches the top without losing a step'. This didn't dissuade him. He began to wonder, how would NASA fare in the future if Seamans weren't there, doing his excellent job? What if he, Webb, wasn't there to defend it in Congress, and instead some other person was charged with that task? He wanted to create an organisation that could continue to achieve success in the tougher times ahead, regardless of the individuals running it at the top. It was the ultimate bureaucrat's fantasy, and Seamans wasn't buying it. 'At that time, Jim was head over heels into the organization thing, where he was much too preoccupied with management affairs. He didn't feel his legacy was the way he wanted it to be. He entered into a sort of mysterious phase, and all of a sudden I felt out of touch.'

Towards the end of 1966, Webb pressed for a centralised secretariat in headquarters that would channel all communications between field centres and programme office chiefs. His hope was that NASA's internal operations could be protected from the hazards of human failure. The rivalries and clashing personalities of individual programme chiefs and centre directors would be mediated by the impartial work of a bureaucracy. The biases of individual personality would be smoothed out by an adherence to procedure, and when Webb at last walked away from NASA he would leave behind him a precision-crafted clerical machine that continued to operate effectively in his absence. Seamans 'didn't believe that clerks within a secretariat would understand the engineering issues sufficiently to make proper decisions about who needed to know what, or who should talk to whom. Also, a secretariat doesn't just pass along information. Sometimes it decides, maybe wrongly, not to pass it on. And that's when people get left out of the loop when maybe they shouldn't be.' Webb also insisted that Seamans spend time defining job descriptions for all NASA's senior people so that precise lines of accountability and responsibility could be drawn up: yet another of the already dizzying 'organisational charts' which were supposed to help tidy NASA's interior life. It seemed a complete waste of time, and almost certainly diverted the energies of Seamans and other key staff at the very moment when – had they but known it – they really needed to be paying special attention to certain problems within Apollo.

As Webb went on the lecture trail to universities, private corporations and chambers of commerce around the country, selling his grand ideas of space age management to anyone who'd listen, his gravest error was to be seduced by dreams while losing touch with hard metallic engineering realities on the ground.

Fault lines

By 1965, problems were emerging between von Braun's people at the Marshall Center in Huntsville and the North American Aviation team working on the Saturn's hydrogen-fuelled upper stage, the so-called S-II. In September, a trial version of the stage failed an important test at the North American factory. Its hydrogen fuel tank was filled with water at a pressure 50 per cent greater than it would normally accommodate. This margin had been built into the design, but only in the knowledge that the tank's aluminium skin was frozen into extra hardness by contact with lightweight super-cold liquid hydrogen, not heavyweight room-temperature water. The tank burst apart with a force that could have caused serious injury if anyone had been in the way. Von Braun's deputy Eberhard Rees was furious. He put together a special group of trusted Marshall insiders to investigate, giving them the acidly judgemental title, 'S-II Catastrophic Failure Evaluation Team'. John McCarthy, North American's unfortunate man on the scene, defended his company as best he could. 'We designed it to a hundred and fifty per cent and it broke at a hundred and forty-four. What the hell do you guys want?' Von Braun was a little more sympathetic, although he had a tendency to strip apart and check every last piece of hardware that contractors delivered to Marshall. Webb told him: 'You have to trust American industry', and the practice was discontinued. Nevertheless, the conservative German rocket experts spoke a different language from the Californian metal-benders, and tensions continued to escalate.

Meanwhile, NASA's Joe Shea, the hard-driving Bronx-born systems engineer appointed by George Mueller to oversee progress on the Apollo command module, was reporting his unhappiness with North American. A devout Catholic, he had a philosophical and spiritual attachment to Apollo that often made him an intense and unapproachable character to work with. His former career as chief systems engineer on the Minuteman missile had been exemplary, yet he somehow lacked the skill to deal with blue-collar technicians and metal-benders on the shop floor of a factory. He decided that one man in particular, Harrison Storms, was the root of all the problems. Seamans was inclined to agree. 'We were concerned about him. He was a very hyper personality, and we weren't sure that he was as thoughtful as he should have been when NASA's concerns were brought up about the schedules, the dollars, the workmanship and so on. And over him, there was really only one other person, and that was Lee Atwood. And Lee was a very fine person. At that time he seemed to be a little on the old side to most of us ...'

George Mueller decided to apply some pressure. In the autumn of 1965, General Sam Phillips, an intelligent and painstaking engineer-manager on loan to Apollo from the Air Force, was dispatched with an aggressive 'Tiger Team' of inspectors to swarm over every last inch of North American's space division and report back on its failings. The aim was to scare the company into better performance.

Over-reaching

Webb gained confidence from NASA's successes in orbit. On 18 July 1966, John Young and Michael Collins successfully docked Gemini 10 with its unmanned Agena target vehicle and used the Agena's engine to boost their combination craft into a higher orbit. Then, in an accomplishment straight out of science fiction, they flew to the old Agena from Armstrong and Scott's spin-dizzying Gemini 8 mission, which had been patiently and blamelessly orbiting since March. Collins space-walked across and retrieved a panel designed to test how different materials stood up to long-term exposure to space. Two months later, Gemini 11, with Pete Conrad and Dave Gordon aboard, docked with their Agena and achieved a giddy 860-mile-high orbit, again using Agena's engine.

Jim Lovell and Buzz Aldrin closed the programme with Gemini 12 on 11 November 1966. There was a problem with the radar that was supposed to track the distance between their capsule and the waiting Agena, but Aldrin wasn't fazed. He'd written a major thesis on orbital rendezvous to qualify as an astronaut. With pencil and paper, he worked out what thrusters to fire for the docking. He subsequently spent five hours space-walking, all the while keeping his heart rate down to a modest 120 beats per minute. He couldn't have seemed cooler about the whole thing, especially since the hand-rails on the Agena's skin, installed at his insistence, allowed him to control the vertiginous tumbling that had exhausted other space-walkers before him. NASA's astronauts had been selected and trained to be as reliable as machines, while the machines they rode could be as temperamental as living creatures. Between them, Man and Machine forged a union of 'can-do' excellence which impressed much of the world, and secured Webb in his conviction that perfect management was the key to it all: a system which, surely, could make managers on the ground as reliable as the astronauts, and as interchangeable.

Webb's ideal was certainly reflected in the way NASA presented itself to the broader culture. He had tried to de-emphasise the astronauts' prominence while promoting the collective strengths of the Agency as a whole. By the late 1960s, the liberal journalist Norman Mailer found that:

> *Everybody at NASA was courteous, helpful, generous of information, saintly at repeating the same information a hundred times, and subtly proud of their ability to serve interchangeably for one another, as if the real secret of their discipline and their strength and their sense of morale was that they had depersonalized themselves.*

Glossy public relations disguised some senior NASA staffers' concerns about their chief. In November 1966, a few days after Gemini 12's successful conclusion, Webb gave a week-long series of lectures at Princeton, where his theories of large-scale space age management left some of his audiences bemused. His determination to impose management changes had become a serious headache for Bob Seamans, who tried politely to voice objections. 'He said he was very troubled about my views on organization, because he was not sure that I fully comprehended the problems of running a major organization in the government.' Webb called for exhaustive briefing papers to be used as a guide to the changes. Seamans was depressed by these demands. 'To my mind, this was just another diversion from our major responsibilities.'

As the technical staff in the field centres and on the shop floors of countless contractors pushed to complete Apollo, Webb's more grandiose ideas seemed increasingly irrelevant. His brilliance as a politician and administrator was not necessarily matched by a similarly exacting grasp of engineering detail. Apollo, in its last laps, was in the hands of practical-minded makers and doers now, and hard-pressed ones at that. Ensuring that the amps pulled by the heaters in an oxygen tank properly matched the supply coming out of the fuel cells and storage batteries was more urgent, in most minds, than defining the job titles of the people who made those checks and signed off on the paperwork. And so, a number of people within NASA began to field less day-to-day information to Webb, because life was made easier that way. Choosing his words with care, historian John Logsdon identifies a dangerous lack of candid communication during the latter half of 1966. 'More and more, there was a discrepancy between what Webb thought was happening within NASA, and what was actually happening.'

By the Christmas of 1966, the first supposedly flightworthy Apollo spacecraft had come out of the North American factory and had been delivered to the Florida launch site, where it was mated to a Saturn 1B booster, a pint-sized although still powerful member of the Saturn family. A long series of checks and tests began, as the gleaming white spire of the rocket stood upright alongside its red gantry. In Washington, Webb felt sufficiently confident to press President Johnson about what should happen after the moon landings. Was the Apollo hardware simply to be abandoned, or could it be adapted for different and perhaps longer-term adventures in space? The White House had no firm answer, but it looked as if an additional $455 million in funding for the next year, 1968, might be approved so that NASA could at least prepare the groundwork for some kind of an 'Apollo Applications Program'. The plan called for a large space station capable of keeping men aloft for a year at a stretch. The future of the powerful technology NASA had invested in seemed reasonably assured. Events over the coming weeks would imperil everything, even before Apollo had made its first flight.

In the early evening of 27 January 1967, mission commander Virgil Grissom and his crewmates, Edward White and Roger Chaffee, clambered into the first flight-ready Apollo command module, mounted atop its Saturn 1B carrier rocket. This was to be a routine checkout procedure. They would run a simulated countdown with all systems running, but they wouldn't actually ignite the Saturn's engines for take-off. There wouldn't even be any fuel in the rocket's tanks. The goal, for now, was simply to check out the Apollo's internal systems.

As astronaut Chaffee climbed through the hatch, he complained that the capsule's interior smelled of sour milk. Then the radio links glitched. Two ground controllers, one nearby at the Kennedy launch centre and another at mission control in Houston, were having trouble hooking up.

'Do you copy?' asked one controller.

'No, I didn't read you at all. I can't read you. You want to try the phone?'

Inside the capsule, Grissom was furious. 'How are we going to get to the moon if we can't talk between two or three buildings? I can't hear a thing you're saying. Jesus Christ!'

'Can you guys talk together up there in the command module?'

'I said, how are we going to get to the moon if we can't talk between two or three buildings?'

The mood was tetchy as pad technicians locked the Apollo's heavy hatch into place, sealing the crew inside, yet no one was particularly concerned about safety that night. Joe Shea was on the launch gantry to watch the tests at close hand. Grissom wanted him to climb into the capsule and lie on the floor behind the astronauts' couches so that he could get a better sense of their frustrations with the radio links, but in the end everyone decided that the hassle of wiring up a fourth headset and microphone in a three-seater capsule at such short notice probably wouldn't help matters. Shea decided to remain outside the capsule.

Five hours into the test, just after 11:30 at night, Grissom's garbled voice on the crackling radio link said: 'We've got a fire in the capsule.' A few seconds later another voice, possibly White's, was more urgent. 'Hey, we're burning up in here!' There was a brief yell, and then just a hiss of static as the radio went dead.

Suddenly the side of the capsule split open. There was a horrifying 'whoosh!' as the top of the gantry was engulfed in thick, acrid smoke and flames. The pad crew, high atop the launch gantry, tried desperately to get the astronauts out. The smoke was impenetrable, and the heat overpowering. It took four minutes to open the Apollo's hatch. By now, the three astronauts were dead.

PART FOUR
Trial by Fire

Our generation has been taught not to regard NASA as perfect. The Challenger and Columbia shuttle disasters have reinforced the bitter knowledge that space flight is not the flawless occupation of angels and gods. It is merely a human endeavour. Yet there was a time when much of the world believed NASA to be invincible. The Apollo 1 fire* of 1967 represented a great crisis of confidence for all of America. Three decades later we look back on the triumph of Apollo 11's lunar landing, the daring mid-space rescue of Apollo 13 and the seemingly repetitive efficiencies of Apollos 14, 15, 16 and 17 and their hazard-free trips to the moon. We forget that Apollo was nearly stopped dead in its tracks even before the first mission flew. The fire gave ammunition to critics of the lunar adventure at a time when their cost-cutting zeal was beginning to find a willing audience. Webb's greatest task as the chief of NASA would not be just to administer Apollo but to *save* it. His task was made all the more difficult by unfortunate coincidences and associations in his business life, and in his handling of the Apollo contract award.

His first move after the fire infuriated Apollo's enemies. He decided that NASA should begin clearing up its own house well ahead of any external public investigations. He had no intention of hiding anything as far as the fire was concerned. But he was anxious not to allow over-dramatic half-truths to emerge before a full and proper technical inquiry had been conducted. No matter what had caused the

* Documentation about the fire often refers to the 'Apollo 204' accident. '204' was NASA's official designation for the capsule during its ground testing phase. In honour of the astronauts who lost their lives, the designation 'Apollo 1' is more often used today. Had they lived, Grissom, White and Chaffee would indeed have flown the first Apollo mission into space.

fire, or who was at fault, NASA could not become the target for ravening wolves intent on an easy kill. He knew it was pointless, both from an engineering level and in terms of public perception, to allow anyone within the Apollo team to lead any investigation. He turned, instead, to Floyd Thompson, director of the old Langley Center. Thompson's people worked on aeronautical problems, one of the important but lesser-publicised aspects of NASA's responsibilities. Langley's contact with Apollo was minimal now that Gilruth and his team were installed in Houston. However, Thompson was still an engineer, so he would be well placed to investigate Apollo's failings.

Having determined this course, Webb met as soon as possible with President Johnson. 'Look, we've got to find out what happened, fix it, and be able to fly again. NASA can do that better than anyone else. But you are entitled to have any kind of investigation you want. I'll cooperate with any kind of investigation.'

'All right, I want you to do it.'

'There's just one condition, and that is, you are entitled to change your mind. The President must never be in a position where he can't change on this kind of thing. The only thing I specify is that if you do change your mind, tell me first, so I can handle myself properly.'

Johnson agreed and the two men shook hands. Webb knew that the President wouldn't be able to hold out for much longer against calls for a wider public inquiry, but he had bought NASA a little time to get its facts straight. Above all, he made sure that the dreadful wreckage of the spacecraft wasn't torn apart by hasty and ill-informed outsiders with axes to grind. The slow and forensic disassembly of the evidence had to be conducted by a team of NASA's experts, even if they were not drawn from within the Apollo programme itself. Once they had gathered their purely technical evidence, Webb knew that a wider political, financial and moral *verdict* on their findings most probably would be delivered by Congress. This subtle distinction was lost on many people at the time, who thought that NASA was trying to investigate itself in order to avoid external scrutiny altogether.

If Bob Kerr had still been in charge of the Senate Space Committee, NASA might have been able to pursue Webb's desired course of a full, searingly honest yet still broadly *internal* investigation, followed by responsible and well-ordered interrogations in Congress. But Kerr was long gone, and Senator Clinton Anderson was now chairing the Senate Space Committee. A senior Democrat, Anderson had served on the Atomic Energy Commission, where he'd been favourably inclined towards Kerr-McGee's nuclear power station projects, at one time even urging the Commission to purchase every last ounce of uranium which the company could produce. Anderson was a great supporter of space flight, and he should have been a natural Webb ally. Unfortunately he was bullish, determined, testy, unwilling to give ground: much like Webb himself. The two men tended not to gel, especially given that Anderson had always wanted more influence over space affairs than Webb would allow him.

Back in 1964, the gigantic Nova rocket, a relic from the 'direct ascent' lunar plan, was discreetly cancelled inside NASA. It had never progressed much beyond engineering studies and paperwork, and in the wake of the 'lunar orbit rendezvous' decision, there was little likelihood of any payload being developed to justify its tremendous size and cost. Webb had felt it legitimate to reallocate $400 million out of the old Nova budget and direct it towards Apollo's Saturn V instead. From his point of view, the money belonged to the lunar programme no matter what size the booster ended up or what name it was given. However, the new chairman of the Senate Space Committee had thought otherwise. 'Anderson let it be known in several quarters that I'd better look out. Senator Kerr had been my friend, Senator Kerr would have supported anything, but a lot of people were coming to *him* now – I'd better be careful, I'd better look out.'

The power alliances had shifted. The Committee held three days of hearings about the Nova, during which Webb was able to say: 'The law gave me the authority and I transferred [the $400 million] to the right place.' Anderson and his colleagues admitted defeat but ruled that Webb shouldn't be granted such sweeping fiscal authority in the future. As the last session came to a close, Anderson sidled up to Webb. 'Well, now, you've done all right. You're going to get along just fine. We've known each other a long time and we're going to get along just fine.'

'I'm sure glad to hear you say that, Senator, because several of my friends have said you told them you were going to cut my throat if you could. I didn't believe it, Senator, because we've been friends such a long time.'

'Well, you're getting along all right.'

This chilly exchange was as much in the style of the Cosa Nostra as Congress. Unafraid of intimidation, Webb could handle a knife when he had to. 'I just put him on notice, you see, that I could sting too. But that's the kind of thing you've got to do sometimes, with people who want to do business that way.' Three years had passed since that encounter, and now the Apollo fire was leaving Webb wide open to attack, and his blade was no longer so sharp against Anderson's. He had to prove he could still be 'counted as a grown man, because you can't sort of weasel along on the good will of some powerful Senator'. Maybe not, but there must have been many times during the Apollo 1 crisis when he would have welcomed the support of someone like Kerr. Instead, a hostile Anderson acceded readily enough to pressure from the media, and from other voices within Congress, for a formal investigation of NASA's conduct in the full glare of publicity. An opening session on the Hill was scheduled for the end of February 1967, after which NASA would be allowed a few weeks' reprieve so that Floyd Thompson's Accident Review Board could complete its work. Then the real interrogations would begin.

One young pup decided to jump the gun. Democrats of a certain age will recall Walter Frederick Mondale as Jimmy Carter's Vice President, and later as the decent but essentially doomed candidate who ran against Ronald Reagan for the White House, only to lose by a landslide. Internationalists celebrate Mondale's subsequent return to form as a wise elder statesman, especially in his role as US Ambassador to Japan for the Clinton administration in the mid-1990s. Apollo veterans, on the other hand, remember him somewhat less fondly as an aggressive young senator from Minnesota determined to make a name for himself at NASA's expense. And they know exactly the day he fired his opening shots: 27 February 1967, at the opening session for the hearings. He was out for blood, and Jim Webb – a fellow Democrat, of all people – was his chosen victim.

The bitterness is still apparent today. An Emmy award-winning TV series charting NASA's history aired on the HBO Channel in 1997. *From The Earth to the Moon* was a finely crafted dramatisation of the Apollo lunar effort, produced by ardent space fan Tom Hanks at a cost of more than $68 million. One episode covered the Apollo 1 fire and the Senate hearings that followed. A key moment shows a hurried private conversation between Webb and Mondale in a corridor just after a particularly savage session in the main chamber:

'Do you really want to kill Apollo?'

'I'm sorry Mr Webb, but I've got a job to do, and I'm going to do it.'

'With all due humility Senator, what did we do wrong?'

'Well that's what I'm going to find out.'

'No, I mean, why are you so down on us? You and I are both Democrats. Going to the moon was Kennedy's dream.'

'It was *one* of his dreams. Jack Kennedy had a lot of dreams.'

The characterisation and casting of Mondale strongly suggests that he viewed Apollo as a waste of national resources, and NASA as an over-sized agency that needed its wings clipped. This semi-fictionalised account reinforced (among modern space fans at least) the more or less historically accepted idea that Mondale was an enemy of space flight in general. Quite by chance, he heard that he'd been cast as the dark villain of the HBO film. 'I was deeply hurt. I didn't have the stomach to look at it. But somebody came up to me after they saw the show, this Tom Hanks thing, and they said, "They just made a monster out of you!" I don't know what explains this virulence. Here we are, and it's more than thirty years later, but I apparently touched some pretty raw nerves.'

The HBO writers had their own sources for the terse exchanges in their script. Webb's chief legal counsel Paul Dembling recalled: 'Mondale tore around the Hill saying to anyone who'd listen, "Webb's lying. He's a liar, and I'm gonna get him." I don't know why, but he really had it in for us.'

According to Mondale, what actually happened was that a journalist approached him and said he'd heard that General Sam Phillips, a senior Apollo manager at NASA, had written a report savagely criticising North American. The report apparently implied that North American should be fired altogether, they were *that* bad. Mondale thought this sounded pretty serious, and he decided to ask some questions at the hearings. It's not that he had anything against Webb or NASA personally back in 1967, he insists, 'but I did think they were trying to cover up some big problems in the space program. I knew they had to be discussing these things privately, and I wanted to see if they would bring them out into the open. They didn't want to discuss it in public, so I pressed them.'

There were grumblings at the lower levels of the space hierarchy too. Thomas Baron, a safety inspector for North American, had compiled a 50-page list of grievances about Apollo, which he leaked to the press almost immediately after the fire. In his opening essay he wrote:

> *It has often been said that 'People must do what they think is right.' In many cases this has been a costly quotation to follow, but it is probably one of the very few ways we have of advancing ourselves as a nation. There are too many opportunities for organizations to live off of the taxpayer. It always seems that the more tax moneys that can be had, the more this money is wasted ... In my opinion, North American Aviation has had the funds to correctly administer a Space Program without compromising the safety of its employees, the astronauts, or the objectives of the Project itself ... North American Aviation has not, in many ways, met their contractual obligations to the United States Government or the taxpayer. I do not have all the information I need ... Someone had to make known to the public and the government what infractions are taking place. I am attempting to do that. Someone else will have to try to correct the infractions.*

1. James Edwin Webb, NASA Administrator from February 1961 to October 1968. (photo: NASA)

2. Senator Robert S. Kerr. 'I represent myself first, Oklahoma second, and the United States third, and don't you forget it!' (photo: Kerr Foundation)

3. Hugh L. Dryden, the wise elder counsel to NASA in its earliest years. (photo: NASA)

4. Robert Seamans, Wernher von Braun and President Kennedy at the Florida launch centre on 16 November 1963. Kennedy was assassinated one week later. (photo: NASA)

5. Apollo's headstrong 'Czar', Brainerd Holmes, consults with Wernher von Braun in November 1961, with systems expert Nicholas Golovin standing behind them. (photo: NASA)

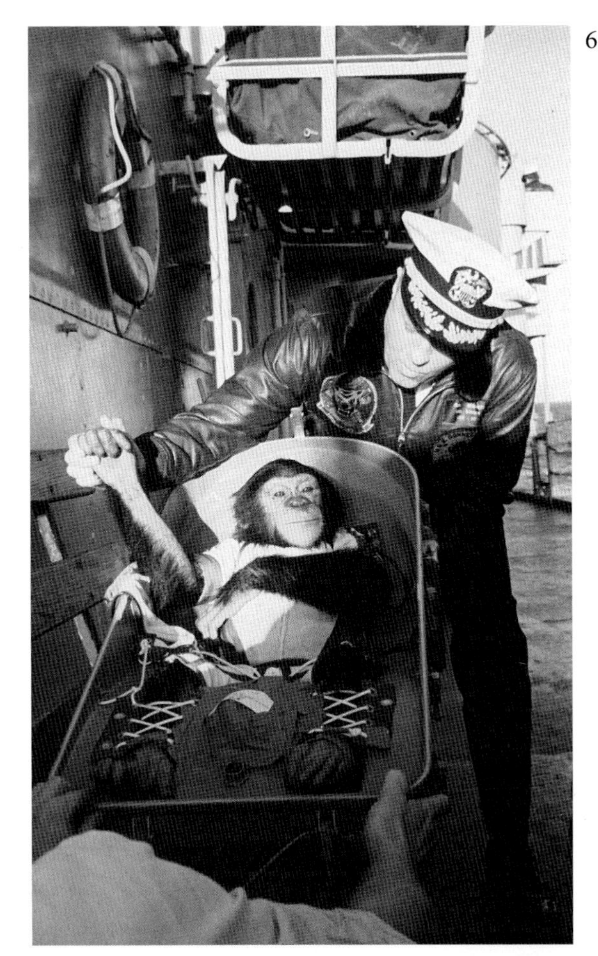

6. Space chimpanzee Ham is recovered after his splashdown at the end of his uncomfortable Mercury flight. If his ride had been a little smoother, history would have worked out very differently. (photo: NASA)

7. North American Aviation's X-15 experimental rocket plane swayed NASA's decision to award that company the main Apollo spacecraft contract. (photo: NASA)

8. Dr George Mueller gives a Saturn V orientation talk to President Kennedy and officials in a blockhouse at NASA's launch centre. In the front row, from left to right, are: George Low, Kurt Debus, Robert Seamans, James Webb, President Kennedy, Hugh Dryden and Wernher von Braun. (photo: NASA)

9. On 3 June 1965, Edward White became the first American to step outside his Gemini IV spacecraft and drift in the zero gravity of space, attached to his capsule only by an umbilical cable. In January 1967 he was among the crew of Apollo 1. (photo: NASA)

10. President Lyndon Johnson shows off photos of astronaut Ed White during his historic space-walk on the Gemini IV flight in June 1965. From left to right are: Robert Gilruth (background), Ed White, President Johnson, Robert Seamans, Gemini IV commander Jim McDivitt, and James Webb. (photo: NASA)

11. An aerial view of the Manned Spacecraft Center (renamed the Johnson Space Center in 1973) in Houston, Texas, shows the gigantic scale of NASA's operations. (photo: NASA)

12. Robert Gilruth (far right) introduces the Apollo 1 crew during a press conference in Houston. From the left are astronauts Roger Chaffee, Edward White and Virgil 'Gus' Grissom. (photo: NASA)

13. Astronauts Gus Grissom, Ed White and Roger Chaffee pose in front of the launch gantry housing their Saturn 1B launch vehicle. (photo: NASA)

14. The Apollo 1 capsule under construction at North American Aviation's main plant in Downey, California. (photo: JSC)

15. Gus Grissom talks to a support technician in this view of the Apollo 1 capsule's complex interior. (photo: NASA)

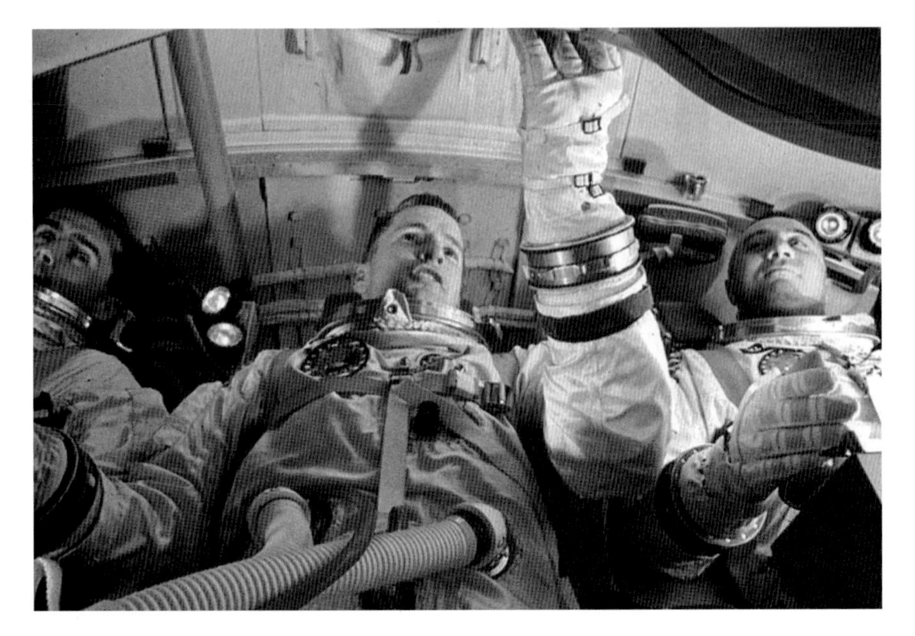

16. The Apollo 1 astronauts seated inside the capsule during earlier training exercises. (photo: NASA)

17. The Apollo 1 Command Module and its rear Service Module are mated together prior to the fatal test. (photo: NASA)

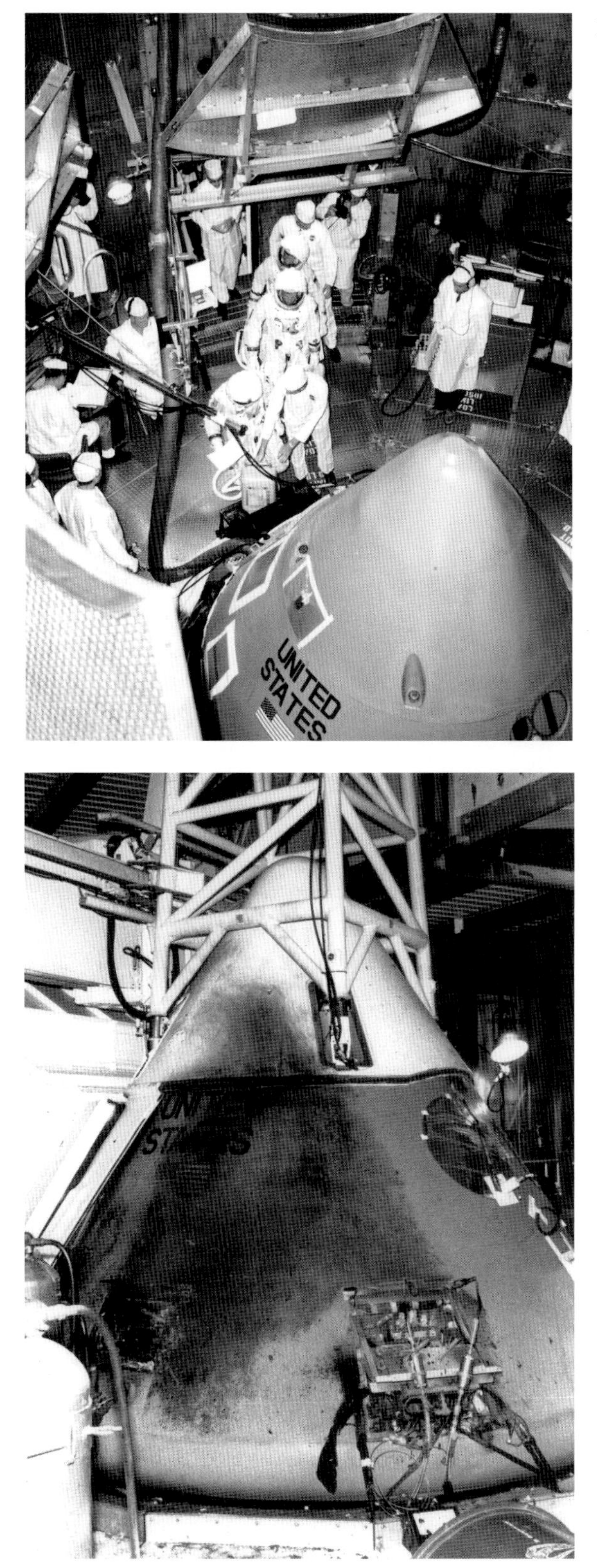

18. The Apollo 1 crew walk across a high gantry towards their flawed capsule on 27 January 1967. This overhead view shows the cramped conditions the astronauts had to endure. (photo: NASA)

19. A view of the charred Apollo 1 capsule. If the emergency escape rockets on the tower at the tip of the capsule had caught fire and exploded, dozens of support technicians working on the launch pad might also have been killed. (photo: NASA)

20. A view through the hatch of Apollo 1 shows the extreme damage caused to the interior of the craft by the sudden fire. (photo: NASA)

21. NASA on trial in Congress. From left to right are: Robert Seamans, James Webb, George Mueller, and General Sam Phillips. (photo: NASA)

22. Lyndon Johnson as Senate Democrat Majority Leader during the last years of the Eisenhower administration, with his trusted aide and confidant Bobby Baker at his side (left). (photo: Time-Life/Getty Images)

23. An early test version of a Saturn V rests inside the giant Vertical Assembly Bay at NASA's Florida launch centre in 1968. Its grandeur impressed the world. Today's machines aim towards sleekness and miniaturisation rather than sheer brute size and power, yet it's a pity that our generation will never have a chance to experience such a mighty rocket. (photo: NASA)

24. Massed ranks of launch control staffers remotely check the readiness of a Saturn V booster at NASA's Florida launch complex. Webb hoped that entire sectors of American society might be run with this kind of efficiency and focus. (photo: NASA)

25. Neil Armstrong, Michael Collins and Buzz Aldrin stand near the Apollo/Saturn space vehicle destined to send them to the moon in July 1969. James Webb had already resigned from NASA, just as it made ready for its greatest triumph. (photo: NASA)

26. Apollo 11 taking off at the start of the first lunar landing mission. It was at once an epic achievement and a historical dead-end. The landing missions were terminated just three years later. (photo: NASA)

27. A mournful ex-president Lyndon Johnson watches the launch of Apollo 11 on 16 July 1969, accompanied by Vice President Spiro Agnew. The new president, Richard Nixon, chose to stay away. His advisors suggested that if there were to be any kind of launch failure on this historic day, it might damage him politically. (photo: NASA)

28. From left to right, Wernher von Braun, George Mueller and Sam Phillips celebrate the successful launch of Apollo 11 on 16 July 1969. (photo: NASA)

29. The Grumman Corporation's lunar module was a delicate construction whose teething problems during manufacture were almost as painful as those of North American Aviation's command module. (photo: NASA)

30. Apollo 11 astronaut Buzz Aldrin poses for a photograph beside the United States flag during the first moon walk on 20 July 1969. Was this flag-waving the sum total of the Apollo experiment, or had NASA truly made an evolutionary leap into the future? (photo: NASA)

31. A view of the earth as seen by the Apollo 17 crew travelling towards the moon in December 1972. Images of the earth from deep space were among the Apollo project's most important prizes. (photo: NASA)

32. The Apollo 1 fire was the first great NASA disaster, but not the last. The space shuttle has proved an even more difficult and dangerous machine. This photo shows a section of the shuttle Challenger after it was destroyed in an explosion on 28 January 1986. (photo: NASA)

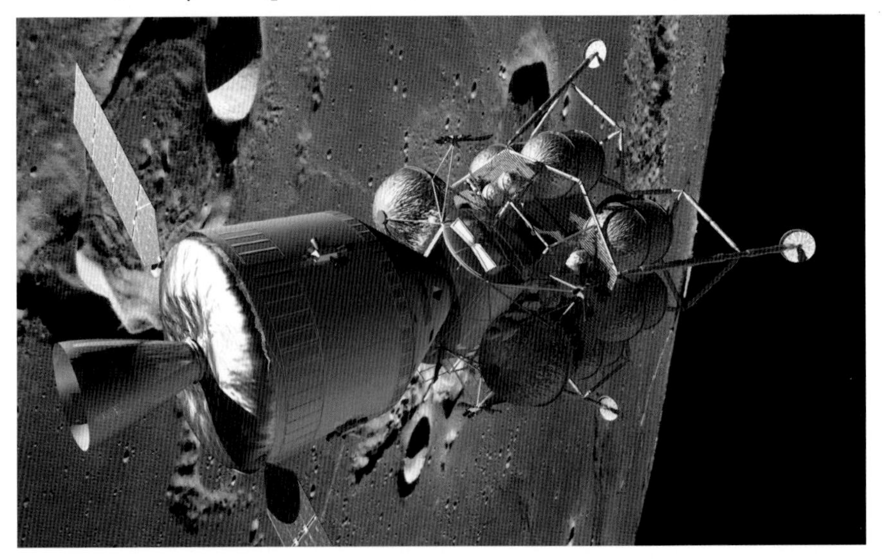

33. NASA's plans for a return to the moon in the coming decade depend on hardware incredibly similar in design to the 1960s Apollo vehicles. (photo: NASA)

26. Apollo 11 taking off at the start of the first lunar landing mission. It was at once an epic achievement and a historical dead-end. The landing missions were terminated just three years later. (photo: NASA)

27. A mournful ex-president Lyndon Johnson watches the launch of Apollo 11 on 16 July 1969, accompanied by Vice President Spiro Agnew. The new president, Richard Nixon, chose to stay away. His advisors suggested that if there were to be any kind of launch failure on this historic day, it might damage him politically. (photo: NASA)

28. From left to right, Wernher von Braun, George Mueller and Sam Phillips celebrate the successful launch of Apollo 11 on 16 July 1969. (photo: NASA)

29. The Grumman Corporation's lunar module was a delicate construction whose teething problems during manufacture were almost as painful as those of North American Aviation's command module. (photo: NASA)

30. Apollo 11 astronaut Buzz Aldrin poses for a photograph beside the United States flag during the first moon walk on 20 July 1969. Was this flag-waving the sum total of the Apollo experiment, or had NASA truly made an evolutionary leap into the future? (photo: NASA)

31. A view of the earth as seen by the Apollo 17 crew travelling towards the moon in December 1972. Images of the earth from deep space were among the Apollo project's most important prizes. (photo: NASA)

32. The Apollo 1 fire was the first great NASA disaster, but not the last. The space shuttle has proved an even more difficult and dangerous machine. This photo shows a section of the shuttle Challenger after it was destroyed in an explosion on 28 January 1986. (photo: NASA)

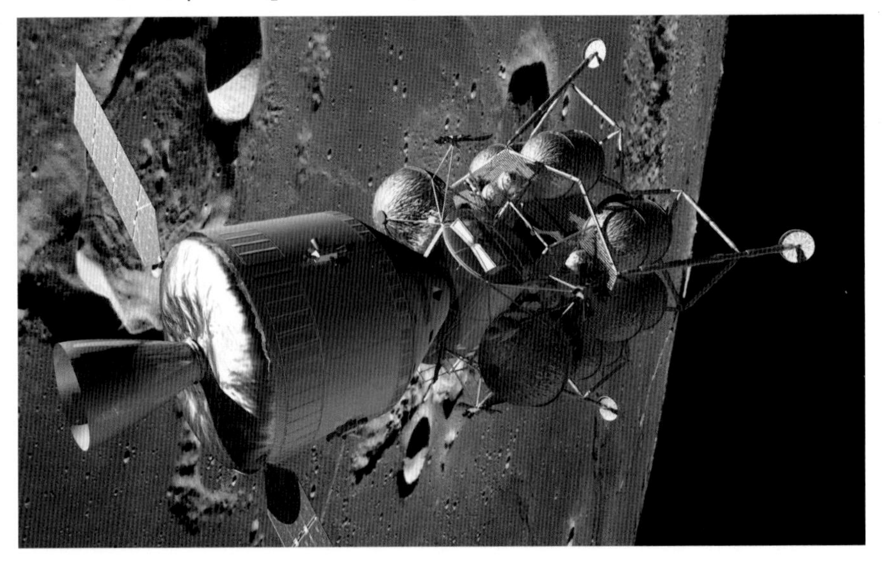

33. NASA's plans for a return to the moon in the coming decade depend on hardware incredibly similar in design to the 1960s Apollo vehicles. (photo: NASA)

NASA on trial

Monday 27 February was the opening day of the hearings: the day when Jim Webb's NASA began to die and the somewhat diminished space agency of post-Apollo times was born. The pivotal moment came after Senator Edward Brooke had finished asking Mueller and Seamans some fairly gentle questions about the service module: the fuel, oxygen and engine compartment that sat astern of the doomed command module. Everyone seemed to agree that it wasn't an issue. It hadn't been seriously damaged by the fire, and no major modifications were required in its design. Brooke then handed the floor to Senator Mondale, who began his questioning with what seemed the friendliest of touches. The two-man Gemini capsules paving the way for Apollo had been, for the most part, 'an astounding success', he acknowledged. He then applauded that safety was now to be the new watchword. 'As much as we would like to beat the Russians to the moon, I am glad to see it deeply imbedded in NASA policy not to let that consideration affect the safety of our men.'

For one tantalising moment there was a sense that the day's business was coming to a close with relatively few wounds inflicted, and the young Mondale was merely injecting his voice onto the record. Just so that he could say he was there, that he'd taken part, he'd nailed his name onto the pages of history during this nationally important inquiry. And then, seemingly from out of nowhere, came the subtle priming of his explosive charge.

'Is there anything to be learned in comparing the successful operation of the Gemini series? And let me preface this by saying that I have been told – and I would like to have this set straight if I'm wrong – that there was a report prepared for NASA by General Phillips, completed in mid-1965 which very seriously criticized the Apollo program for multi-million dollar overruns and very serious inadequacies in terms of quality control. Would you be able to comment on that? Is there such a thing as the Phillips report?'

The entire room fell silent for a moment.

Mueller took the question first, and dissembled skilfully. His answer sounded technical, logical and cool, as befitted the man, but it was truly an attempt to skate over the subject with a blizzard of words. 'The Apollo was, in fact, a better-quality article than the first Gemini spacecraft', he explained, so you had to expect arguments between NASA and its contractors on a fairly regular basis. North American Aviation was no better and no worse than any other company. Yes, General Phillips, a highly experienced senior figure in the Apollo management team, had prepared plenty of reports on plenty of wayward contractors in his time, and 'not all of them have been complimentary … I think you will recognize that in any research and development program, we have to pay attention to areas of weakness, both within NASA and among the contractors.'

But Mondale persisted. 'Is it your testimony that there was no such unusual Phillips report? Is that rumor unfounded?'

'I don't know of any such unusual report', Mueller replied.

'Was there a report in which General Phillips recommended looking for a second source [an alternative company to build Apollo]?'

'I do not recall such a report.'

'So are you saying that what has been reported to me, you regard to be inaccurate?'

Seamans was becoming anxious that Mueller's denials were becoming too emphatic, too absolute. 'Mondale was the junior senator, so he came on last. And it had been a long, long session, and so by the time he came along – I won't say we were waning, but we'd been through quite a bit already. And suddenly he said, "Was there a report that had been written about North American that indicated they weren't doing a good job?" And I had the sense that Mondale wasn't just on a fishing trip. He knew something, and the question was – *what* did he know? And so I introduced the thought – I responded to Mondale that we did indeed periodically carry out extensive reviews of our contractors, and was he perhaps referring to the results of one of those reviews?'

Actually, the transcripts reveal that Seamans made a serious error. He'd missed the nuance of what Mueller had been trying to suggest a moment earlier: that critical contractor reports were made routinely, in connection with a wide gamut of space programmes, and none of them were any more significant than any other. At one point, Seamans came dangerously close to discussing the Apollo contract in particular. 'We do look at the performance of each one of our contractors, and I can remember discussing the possibility of a second source on parts of the Apollo', he said. 'This was not directed at having the work go to [another company] but we do happen to believe that competition is a healthy thing, and from time to time we do review these possibilities.'

Webb interjected. 'I want to add one other thing. There is nothing in the preliminary findings of the [Review] Board that points to the cause of the accident or fire as being generated by any contractor.'

Mondale took stock while other members of the Committee asked about ignition sources in the capsule, but he soon returned to his favourite theme of the day. 'What I'm getting at is this allegation – this serious criticism – made about this so-called Phillips report.'

Webb sensed terrible danger. 'Senator Mondale, are you thinking of a report circulated by a former employee of the North American Aviation company recently, within the last six weeks?' He was alluding to the Baron business, but he would have known there could be no confusing the names 'Thomas Baron' and 'Sam Phillips'. The purpose of his interruption was to halt his NASA colleagues' testimony before things could get any worse.

'No, the so-called Phillips report, which came out in mid or late 1965, conducted by NASA on the Apollo program', Mondale intoned relentlessly.

'Let us look it up', Webb said.

Mondale eased back on the throttle, for that day at least. 'Webb looked dazed and stunned. I don't know if he knew about the Phillips report specifically, but to this day, I think he must have done. Anyway, from that point on, it became inevitable that they'd have to produce it.'

As soon as everyone rose from their seats and prepared to quit the chamber, Webb pulled Seamans aside. 'I want you to come with me.' His face was rigid with fury. Outside, they climbed into Webb's car, along with legal counsel Paul Dembling. Webb immediately cranked up the sound-proofed window that separated passengers from the driver in front. Then he rounded on Seamans. 'There's no excuse for volunteering information like that! We're dealing with matters that could result in millions of dollars of lawsuits. This is not the kind of friendly technical inquiry you're used to. And what's this report Mondale's talking about? Why haven't I been told about it?'

Seamans was appalled. 'In all the time I worked with Mr Webb, I met with him every single day. Before we'd go home I'd tell him about all the horrors, all the problems of the day, and he'd usually say to me, "Well you've gotta admit Bob, this is an interesting business." And for the life of me, I can't believe that I didn't keep him informed about everything. But he claimed after the fire that no one had ever told him about these problems. And that's one of the things I feel badly about – that somehow he felt he'd been kept out of the loop.'

Webb had spent the last few weeks defending his Apollo managers against charges of incompetence, but privately he felt let down that his closest colleagues hadn't reported to him on the issues of fire safety that had led to the disaster in the first place. Now, here was this 'Phillips Report' which was obviously of such great interest to Mondale, yet which no one within NASA had taken the trouble to warn him about. The impression that Webb had given at the hearings was that he was trying to hide the report, but he really hadn't known precisely what Mondale was referring to. He had been made to appear shifty and evasive, and he was furious.

Dembling, sitting beside Webb and Seamans in the car as it pulled away from the Hill and back towards NASA headquarters, was experiencing a crisis of conscience on his own account. 'That very morning, I had prepared to go up to the hearings. At the last minute, as I was walking out, General Phillips' assistant came in and said, "I have some documents that I'd like you to look at. I know you're on your way out, but when you get back, maybe we can talk." I said, "OK, I'll take a look as soon as I get back, I don't have time now. I have to meet Jim Webb in a couple of minutes." And in the car, after Mondale had done with us, I already knew what I'd find when I got back to my office. And sure enough, there's this file called "Phillips Report" on my desk, and I hadn't had time to look at it.'

He told Seamans what he'd found. Seamans didn't have the stomach for what had to happen next. Dembling bravely volunteered to walk the Phillips material into Webb's office. 'He said, "Get me those guys. And get 'em the hell up here real fast." He didn't like to be one to be accused. He felt he always had very good relations with Congress. He never wanted to be put in a position where he looked like he was obstructing what they were seeking, or not giving them full information.'

Webb now grilled Phillips, Seamans and Mueller about the investigative 'Tiger Team' they had sent into North American in November 1965. On 15 December of that same year, Phillips delivered a 'set of notes' to Mueller, listing the Tiger Team's uncomfortable findings. This was the so-called 'Phillips Report' that Mondale had unearthed. And it did indeed include some stunning paragraphs and phrases:

> *NAA's [North American Aviation's] inability to meet spacecraft delivery dates has caused rescheduling of the total Apollo program. NASA has been forced to accept slippages in key milestone accomplishments, degradation in hardware performance, and increasing costs ... Even with due consideration of hopeful signs, I could not find a substantive basis for confidence in future performance ... There is little confidence that NAA will meet its commitments within the funds available for the Apollo program.*

Also filed with the report was a letter from Mueller to North American's chief executive Lee Atwood.

> *I submit that the record of this program makes it clear that a good job has not been done ... I had hoped that a letter such as this would not be necessary. However, I consider the present situation to be intolerable and can only conclude that drastic action is in the best national interest.*

Mueller hadn't specifically discussed this letter with Webb until now. As far as he'd been concerned, bullying North American into better performance really had been a matter for Apollo management to deal with, and he'd seen no obvious need to involve Webb personally. Mueller had assumed that an internal report, generated for internal purposes, would stay internal. 'I guess I was surprised, and Jim Webb was surprised when the report began to surface, because we had thought – I had thought it was carefully buried and destroyed. Not the knowledge, but the report itself. But obviously, once you create a report like that, you can never really destroy it. Everybody has a secret copy.' Webb knew better than Mueller how these things worked. His mood darkened yet further when ABC news reporter Jules Bergman broadcast a detailed story about the report, which he had obviously researched well ahead of that day's hearings. NASA had been comprehensively ambushed. Bergman had somehow encountered a copy of the report two weeks earlier. He was the mysterious source who handed a copy to Mondale. Webb could only conclude that someone at NASA had leaked the report, and he felt personally betrayed.

Even today, Seamans is at a loss to explain exactly how such a significant thing as the Phillips Report had entirely escaped Webb's notice until that point. 'I'm sure we'd brought it up at the time.' Seamans thinks that he might once have seen a copy of the report with Webb's signature on it – a signature dating from long before the Congressional inquiry. But he's not sure. NASA today has no trace of it. Webb acknowledged that he'd had some discussions with Seamans about North American's laggardly performance, and had been told about punitive revisions to the terms of the company's contracts towards the end of 1966, but 'if he told me about it, it was casual and not flagged as something important'.

Historian John Logsdon has also considered this strange question. 'Mueller was an even stronger character than Brainerd Holmes. Webb had even less control over him. He was running the Office of Manned Space Flight as essentially a separate organization, to the point that Webb didn't always know what he was up to. You have to keep in mind that the Phillips Report was originally made for Mueller to see, not Webb.'

For his part, Mueller didn't even acknowledge the existence of a formal 'report'. It was just a working document, an unbound 'set of notes' stacked loosely together with some covering letters and projection slides. Nevertheless, he knew he'd dropped his boss in at the deep end. Inadvertently he had 'caused a credibility gap, particularly between Jim Webb and Congress'. As Webb pressed his embarrassed colleagues for explanations, Mueller didn't relish being put under the spotlight. 'Jim was a person who wanted quick and straightforward answers to questions and problems. It was a stressful time.'

The incident also damaged the trust between Webb and his immediate deputy. According to Seamans: 'Our relationship had been exemplary. He'd been just a great boss. He'd given me a lot of latitude. I felt I'd kept him informed well, we'd worked well together. And all of a sudden I felt as though he had me sort of under the gun. At first he would bring up things about other people. He'd say, "George Mueller – he's sort of let us down, hasn't he?" And I'd say, "Well you've got to realize what he's accomplished." I tried to explain to him, none of this was really a case of mismanagement on anyone's part. But then, a few people came to me and said, "You must realize, what Webb's saying about George, he's also saying about you." I could tell that our relationship really was changing ... And I feel that that was unnecessary. That he could have dealt with the problem without trying to hang George Mueller or hang me.'

It had finally dawned on Webb that his 'system of management wasn't working. It was eroded ... Shea and Mueller had bypassed Gilruth and were running things on their own ... They were rushing too fast in testing and in [scheduling] early flights. We had trusted them. We had a system laid out that said you don't proceed until every anomaly is found and corrected. This system was not followed.' As for Seamans, Webb believed he had become over-friendly with Shea. What was more, he had committed terrible bureaucratic sins, disbursing 'hundreds of millions [in NASA contracts] without a piece of paper. I had to reconstruct the legal basis for awards already made.' This criticism was more than a little unfair. When interviewed, Seamans tends to gloss over some of the more distressing hours and days of his tenure at NASA. Nevertheless, in his written recollections he cannot avoid commenting on some of the tensions that emerged as Webb cast around for someone to blame:

> *I'm not a psychiatrist but I would say that the fire came as a terrible blow to him. Before the accident, he and his program were riding high. He was front and center, getting acclaim from many, many quarters and deserving it ... Now his house of cards was down. How? Why? Who had made the mistake? Who had destroyed his dream? It was necessary, of course, to carry out a complete and careful investigation ... but Jim was not interested in investigating the engineering. He wanted to know what individuals had failed him. He felt personally betrayed.*

It wasn't just the fire and the scorching of Apollo. As space historian Roger Launius points out, Webb's deepest ambitions were at risk. 'He had talked about "Space Age management." He had developed all these principles that he was trying to employ in terms of creating the *perfect* organization. He talked about the application of NASA techniques to other sorts of activities such as homelessness and poverty. I mean – he thought this was *portable*. And all of a sudden, in a very fundamental way, this great management system collapsed. And it devastated him personally, and it devastated the people around him.'

Everyone in NASA was stricken with grief, and probably none more so than Joe Shea. He wondered if perhaps he might have been able to spot the fire if only he'd been lying on the floor of the capsule that day. His devotion to Apollo – his spiritual sense of its significance – caused him to take on much of the guilt personally, as if he had committed a terrible sin, and he edged close to breakdown. Given that he was one of Mueller's most significant lieutenants, there was a real danger that Congress would call on him to give testimony. Webb was determined to avoid this, for the sake of both Shea and NASA. He eased an unwilling Shea into a desk job at Washington headquarters, where he could be more easily supervised. George Low, Bob Gilruth's trusted deputy at the Manned Spacecraft Center in Houston, was appointed in Shea's place, charged with fixing the Apollo. Low seemed mild-mannered, almost self-effacing, but like so many of the characters surrounding Apollo, he could be remarkably bold in his decision-making.

Webb then called in some of the last of his dwindling stock of favours from helpful members of Congress, asking them not to subject Shea to a grilling. Whatever the ruthlessness and political point-scoring that Apollo offered to careerists on the Hill, it would have looked bad if any witnesses had been driven to tears by politicians who'd never even seen the Apollo, let alone met and worked with the dead astronauts on a personal basis.

In time, the unimpeachable successes of Apollo would allow Webb, Shea and all their NASA colleagues a more forgiving perspective. For now, the simple truth buried in the subtext of thousands of pages of Accident Reports and Congressional Committee testimonies was this: everyone in general at NASA was to blame, and no one in particular. There was nothing especially wrong with Webb's management system. It was actually very fine. He simply refused to recognise that no such system is perfect. As for his engineers and technical people: overall they had done an excellent job, persuading extremely complicated and hazardous new technologies to function with startling reliability. Even as he tried to find some tangible and therefore fixable flaw in the personnel, the paperwork, the *system*, Webb understood the ghostly elusiveness of the demon that had tortured Apollo: 'A fuelled launch vehicle was regarded as a dangerous and terrifying thing. But [this particular vehicle] was not fuelled. It was not looked upon as a rattlesnake. The tendency was to relax because it wasn't fuelled.'

Air versus oxygen

It wasn't fuelled. The rocket sitting under the capsule, the half-sized variant of the Saturn called the 1-B, was empty. On launch day it would harbour the latent energies of a small thermonuclear bomb, but during this particular capsule test its tanks had not been loaded. There was nothing inside the rocket to explode. But still there had been an explosion ... And it had happened among the supposedly innocent gas inside the capsule that the astronauts actually breathed to stay alive.

We have evolved to live in the atmospheric pressures normally encountered at ground level. Climb up a mountain just a few miles tall, and the rarefying of the air is enough to make you gasp for breath. The higher into the atmosphere you venture unprotected, the less oxygen there is to breathe, and the more the internal pressures of your body cause it to balloon outwards because there's no longer a corresponding pressure of air outside the body to push back against it. At extreme altitudes, your blood vessels burst and your eyeballs bleed – although it's a myth that people explode when exposed to the raw vacuum of space.

High-flying pilots are sealed in a pressurised cockpit, breathing oxygen at a sufficient pressure to keep their bodies and minds in good shape. That's oxygen, not 'air'. Ordinary air consists mainly of nitrogen, a colourless, odourless, non-flammable gas. The oxygen essential to sustain life makes up only a fifth of ordinary air's content. Why, then, do jet pilots breathe oxygen alone and not air? The answer to this riddle can be found in valves and bubbles. A 'two-gas' system inside a cockpit is potentially complicated and prone to producing accidents. There are twice as many tanks, valves and control systems to go wrong. If any imbalance occurs in the ratio of nitrogen to oxygen delivered to the pilots, they may end up breathing too much of the first and too little of the latter. What's more, if a plane is attacked, or suffers some other calamity that breaks the cockpit's seal and causes the pressure to drop, the nitrogen breathed in by the pilots is liable to expand in the bloodstream, causing bubbles. These excruciatingly painful and dangerous expansions, known as the 'bends', are greatly feared by deep-sea divers, who adopt special techniques to avoid them, such as cleansing their bodies of gaseous nitrogen before a dive and making sure to use pure oxygen rather than air in their breathing tanks. In a pure oxygen cabin, the bends can't arise because oxygen is more fully absorbed into the blood than nitrogen. And so, the builders of high-performance aircraft and spacecraft have long favoured oxygen to keep their crews alive.

While air fanned vigorously onto burning logs will stoke the flames, pure oxygen is highly flammable and needs no such encouragement to lend ferocity to a fire. After all, that's why it so commonly serves as the partner to kerosene or liquid hydrogen in the explosive reaction that powers a rocket engine. With this explosive hazard in mind, Apollo's designers took care that the cabin pressure inside the spacecraft would be as low as possible, while at the same time allowing their astronauts to breathe in comfort. In the vacuum of space there was no need to worry about external pressures crushing the capsule. An interior oxygen pressure of just five pounds per square inch was sufficient to keep the astronauts comfortable: just a third of the pressures normally encountered at ground level on earth. The astronauts were no more inconvenienced than mountain climbers on the foothills of Everest. The shell of the capsule was correspondingly lighter than it would have had to be if it contained gas at normal atmospheric pressures. Just as importantly, the ignition hazards of oxygen at five pounds per square inch were quite minor. If a fire did get started somewhere in the cabin, the astronauts would have time to reach for their extinguishers and put it out.

The problem was that the conical shell of the Apollo command module was designed to withstand pressure from the inside, not the outside. To simulate this condition properly on the ground, the pressure inside the cabin had to be raised slightly above fifteen pounds per square inch, because that was the surrounding air pressure at the launch gantry. Similar tests had been run on all NASA's capsules prior to Apollo without incident. Only after the fire did anyone dare to suggest that any high-school student with a basic grasp of chemistry could have told NASA not to use high-pressure oxygen inside a capsule, even if it was still just sitting on the ground. A trivial short-circuit in the environmental control system caused a spark, which in turn set light to a piece of Velcro fabric used by the astronauts to hold pens, checklists and other small items in place on the capsule's walls. Velcro as a product was not in any way at fault. It was simply that no one had predicted the fire risks it posed when ignited in pure oxygen at above-atmospheric pressure.

All these unfortunate truths were apparent by the time Floyd Thompson delivered the final draft of NASA's report into the fire on the weekend of 8 April 1967. Two thousand investigators from a wide range of disciplines had examined every nut and bolt of the spacecraft. The technical errors were understood, even if they still seemed slightly unbelievable. Little had been said, so far, about the management problems within Apollo, both inside NASA and at North American's plants. However, there were stinging references to the quality of the capsule: the untidy wiring, carelessly discarded hand-tools and general lack of attention to detail:

> *The Board found numerous examples in the wiring of poor installation, design and workmanship. An example is shown in Enclosure 27 where a wrench socket was found in the spacecraft. The Board's investigation revealed many deficiencies in design and engineering, manufacture and quality control ...*

As Congress prepared to reconvene the hearings on 10 April, Lee Atwood began to realise that he and Harrison Storms were expected to shoulder much of the blame for the fire. The tone of NASA's Review Board deeply offended him. 'I felt quite strongly that the [implication that] poor workmanship caused the fire was completely wrong ... The company took the blame for something it wasn't responsible for.' Moreover, NASA had been a difficult client to work with. Senior astronauts had stalked the factory floors, demanding changes to the layout of the cabin interior to suit their purposes, while

NASA kept shifting the designs, making life difficult for North American's fabricators as they tried to cut metal and weld or bolt components into place. The client's specification for Apollo, although very detailed, 'was not a coordinated set of instructions to the manufacturers', Atwood insisted. 'It was a continual sort of a rumble of desires of various elements of NASA's technical staff, or [demands for] information about things that had worked and things that hadn't worked. Resolution of arguments among the NASA people themselves in many cases took quite a lot of time.'

For whom the bell tolls

Webb did seek out Mondale for a private meeting, as Mondale readily admits. 'Oh, he didn't like the way things were developing. He criticized me for bringing up the Phillips report. He said I had a duty to go to him privately and not publicly. I said, "You know, Mr Webb, I don't agree with you. I'm a United States Senator, and this is public business. I think I did exactly the right thing and I'm not gonna get in a position where I have to ask permission to ask questions. I'll reserve that right. I'm a Senator."'

Webb recalled Mondale using somewhat less moderate language: 'I think it's rotten and I'm going to blow it out of the water.' Whatever the exact dialogue, it was an uncomfortable encounter. For the first time perhaps in his entire tenure at NASA, Webb had lost control of the politics of a situation. Mondale would be calling the shots from now on, at least when it came to the Phillips report. Like fencers, Webb and Mondale duelled over the minutiae of submitting documents to Congress in much the way that lawyers often argue over points of law rather than the merits of an actual case. There was a lot of discussion in the hearings about the difference between a 'report', formally printed up and put on the record, and (in Mueller's description) a 'set of notes' never actually intended as the last word. Mondale dismissed these arguments as 'a semantic waltz' and insisted that all the material should be made available. Webb countered with an array of legal arguments. He said, quite correctly, that it would be unlawful for private discussions between NASA and North American to be detailed in a public hearing, because they might adversely and unfairly affect North American's stock prices before any verdict had been reached. However, most of the Phillips report's substance was appearing by now in the newspapers. Mondale and his allies knew that once it was in the public domain it would be increasingly hard for Webb to avoid discussing it in the hearings.

This was just the opening round in a series of brutal interrogations in both houses of Congress. On 10 April, the House of Representatives' Subcommittee on NASA Oversight began its 'full and complete review of the Apollo accident, its causes and effects'. Webb knew what lay ahead: a show trial, an appalling public circus complete with ravening lions and sacrificial victims. In his opening statement he pleaded for sanity. It was perhaps the most poetic speech ever made by a NASA representative:

> *If any man wants to ask for whom the Apollo bell tolls, I can tell him. It tolls for him and for me as well as for Grissom, White and Chaffee. It tolls for every astronaut or test pilot who will lose his life in the space-simulated vacuum of a test chamber or the real vacuum of space. It tolls for every astronaut-scientist who will lose his life on some lonely hill on the moon or Mars. It tolls for government and industrial executives and legislators alike. It tolls for an open program continuously evaluated by opinion-makers with little time for sober thought – operating in the brilliant color and brutal glare of a real-time worldwide mass media that moves with the speed of a TV camera from euphoria to exaggerated detail ... [We] have a grave responsibility to work together to purge what is bad in the system we together have created and supported. We have perhaps an even graver responsibility ... to preserve what is good, and what represents still, at this hour, a high point in all mankind's vision.*

Seamans was impressed by Webb's determination and bravery during the hearings – and by his skilful use of bluster. 'He was a very powerful, forceful individual on a man-to-man basis. He came on extremely articulate. He was a very fast-talking person. After one long hearing, one senator asked the Chairman if there was time to ask Webb one more question. The Chairman said, "There's time for the question, but I don't think we'll have time for the answer." When he got hold of a question, he'd always turn it to his advantage. If we were getting a barrage of questions, he knew how to slow things down by extending his discussion.'

Unmoved by poetic appeals or lawyerly manoeuvrings, Congress expressed its dismay about the capsule's pure oxygen environment, the frayed wires and fatally clumsy hatch. 'The level of incompetence and carelessness we've seen here is just unimaginable' was the prevailing opinion. This was profoundly unfair, given that NASA had previously completed sixteen manned missions in two different kinds of spacecraft (Mercury and Gemini) without losing any crewmen. No one at the time seemed to question the ludicrously high standards expected of NASA with regard to three astronauts, or compare them with the apparently less accountable standards expected of the US military, in the same year that 56,000 American soldiers were wounded in action and 9,400 killed in South East Asia on a mission whose management was uncertain, whose purpose was undefined, and whose execution was flawed. NASA's crewmen knew they were in a risky adventure and climbed willingly into their capsules. The soldiers in Vietnam had little or no choice about their exposure to risks. The war could never have withstood the kind of scrutiny now being applied to Webb and his colleagues.

George Mueller certainly felt that Apollo had been unduly targeted for criticism. 'How do you explain to the public at large that there's a certain risk [in space flight], and you've got to accept that risk? We haven't been able to do that in our society. People are quite willing to accept risk for race drivers, for example, and clearly that's a risky operation. But when one of them gets killed, you don't have a Congressional investigation or a Presidential commission looking at it.'

That's not how the national press saw things. Webb felt that unforgiving journalists saw the fire as 'the defacement of an idol. NASA and the astronauts had been built up for a fall'.

NASA's chief was heading for his own fall. On 17 April, Republican Senator Margaret Chase Smith, someone who was actually quite well-disposed towards Webb, asked a perfectly fair question to which he provided a dangerously flawed answer. 'Would you tell us how the spacecraft contractor was chosen? Was it by competitive bid? Was it chosen by an evaluation board? How did it happen?'

'We had a Source Evaluation Board', he replied. He then tried to explain at some length that 200 experts had been involved in vetting the various proposals for Apollo, but Chase Smith cut in. 'May I interrupt you right there? Was North American the first choice of the Source Evaluation Board?'

'Yes. It was the recommended company.'

He had just made the gravest mistake in his career.

Mr Baron

A separate interrogation occurred at the Kennedy Center on 21 April while a delegation of Congressional inquirers inspected the facilities. Thomas Baron, the disenchanted North American employee whose critical report had been leaked to the press almost immediately after the fire, described his background experience in Navy and Air Force inspection regimes before joining North American Aviation's Hound Dog missile programme for two years as a quality assurance inspector. He had worked on Apollo for the last sixteen months, and most recently he'd been part of the launch gantry team. Olin Teague and his House Subcommittee on NASA Oversight pressed him for more details about the accusations in his report.

'You stated that North American had not lived up to their contractual obligations to the Government. Is that correct?' Teague asked.

'I don't feel that they did. No, sir.'

'Mr Baron, did you ever see their contract?'

Well, no. Of course this relatively junior safety inspector hadn't seen it, let alone read the thing. He admitted that his judgement was based on his past experience of standards usually expected from a manufacturer working on any government missile project.

Teague then wanted to know about Baron's sources. 'Either in your tape or in your report – I don't remember which – you said you met a man in a grocery store who worked on the pad and who knew exactly what caused the fire.'

'This is what he said to me. Yes, indeed.'

'Who was that man?'

'Are you going to press me for that name?'

'Yes. We want to ask him about it.'

Baron surrendered the name of his contact, Al Holmburg, a North American electronics technician. Apparently the two had spoken when they happened to meet at a drugstore in Titusville, Florida, in early February, just a few days after the fire. Holmburg was at the Kennedy Center that very moment and was alerted that Baron was throwing his name into the mix. He volunteered to testify even though he hadn't formally been called. Baron, meanwhile, continued with his evidence, which seemed sincere but lacking in specifics. 'I would say basically that we have had extensive problems in safety, in cleaning materials, in items getting in the spacecraft that weren't supposed to be there, the morale of the people, the pressures put on the people by management – these are the things that really indicate that we don't have the proper management that we should have in this program.' He outlined problems with fire extinguisher systems and breathing masks on the pad, missing headsets and so forth: a catalogue of genuine grievances, yet delivered without the mass of solid technical evidence to which the Apollo 1 investigation had by now become thoroughly accustomed.

Congressman James Fulton challenged him impatiently. 'We are listening to your words. The question is whether the committee should take them on their face value. We need to know who you recommend to us to corroborate what you say. Is it a company? Individuals? Engineers? Systems engineers? Who is it? Pad operations people?'

'I'd suggest that you talk to Mr Holmburg', Baron said, playing what appeared to be his only trump card.

'Oh, we clearly will do that', Fulton assured him.

Baron admitted that he'd discussed his concerns with the press: a fact that didn't please his interrogators. As the somewhat brutal session drew to a close, Congressman Ken Hechler laid into him in no uncertain terms. 'I think it is utterly irresponsible for you to come before this committee and attempt to dignify a conversation that you had in a drugstore ... I feel it is unfortunate that this has been brought before the committee.' There were some discussions of Baron's medical record and his treatments for depressive mental states in the past. He then tried to talk about North American employees using alcohol tank cleaner solutions for illicit booze distilleries, but the

Committee didn't seem very interested. When Baron was asked, one more time, for a detailed list of individuals who might confirm his accusations, he raised a point which has fascinated Apollo historians ever since. 'I have sent to the chairman of this Committee a more thorough report which includes all the names.'

'I have all the names', said Fulton, 'but I read them and said to myself, who should we call?'

'No, sir. You are talking about the 55-page report. I am talking about the 500-page report.'

'Your report went to the chairman of the full Committee, not to me. He told me he received it.'

'I would like it to be submitted as a part of the record, the 500-page report.'

'That means printing it', interrupted Congressman John Wydler. 'That's something we should leave to the Committee, something of that length, whether we want to print it as part of the public documents. We can take it as an exhibit. Whether we will print it as part of the public record is something we should decide after we see it.'

Obviously the Committee decided not to print it. No copy of the '500-page' version of Baron's report can be found today, but the transcripts make it clear that the Committee had definitely received something on that scale. Space historian Roger Launius confirms that a major hunt has been conducted throughout the NASA archives in recent years to try to find it, without success. The mysterious document has become something of a touchstone for Apollo conspiracy theorists. In all probability, it didn't shed much more light on North American's problems than has already been made clear. We may never know.

Teague dismissed Baron with lukewarm thanks. 'I think we are through with you. The Board has found some of the things you have said to be true. What you have done has caused North American to search their procedures. Thank you very much.'

Al Holmburg then came in to the room, and Teague got down to business right away. 'Mr Baron has testified, as I am sure you know, that you told him that you knew what caused the accident. Did you ever tell him anything of that nature?'

'No, sir.'

'Will you put your hand down away from your mouth?' Fulton demanded.

'Did you ever discuss the cause of the accident in a drugstore with Mr Baron?' asked Teague.

'No. I talked to him many times in the drugstore, but that's about it ... I bumped into him accidentally almost every time I met him. I told him I shouldn't even be talking with him because of the report he's writing, and he's probably being watched. He gets all his information from anonymous phone calls, people calling him and people dropping him a word here and there. That's what he tells me.'

Connecticut Congressman Emilio Daddario was caught off-guard by Holmburg's sudden arrival. 'What caused you to come here today? We hadn't scheduled you as a witness. I had no idea. In fact, I can't recall that I ever heard your name before today. What brought you here?'

'Well, Mr Baron had brought my name up a couple of times in here, and I thought I should come in here to defend it.' He went on to paint a portrait of Baron as a gossip-monger. 'He did most of the talking about [the accident] and I listened to speculations on that thing. I never made any comments about what caused it, and I never told him exactly what caused it. I was never near the accident when it happened.'

Baron's reputation had been destroyed. His role in the Apollo drama was now at an end. He may have been the victim of tactlessness and unfair treatment at the hands of the investigating committee. Perhaps a more gentle style of questioning would have been appropriate for a man who was quite obviously not up to the demands of such a serious high-profile interrogation. At any rate, just one week after his grilling, Baron and his wife and two children were killed when a train smashed into his car at a railroad crossing. The circumstances were not inconsistent with a suicide.

The context of tragedy

Two days later, a disaster in the Soviet space programme helped put the Apollo fire into its wider perspective. On 23 April 1967, the first of the new Soyuz spacecraft was propped up against the gantry at Baikonur, ready for firing. The archive pre-launch footage shows its unhappy test pilot, Vladimir Komarov, and some very subdued technicians. It's almost as if everybody knew that a bad day was in store for them. And indeed, Komarov hit trouble as soon as he reached orbit. One of the solar power vanes on the rear equipment module of the Soyuz refused to deploy, and his guidance computers ran short of power. Launch of a second Soyuz that was supposed to rendezvous with Komarov was cancelled while ground controllers worked to fix his power problems. After eighteen orbits (26 hours), mission controllers decided to bring the capsule down to earth. Komarov couldn't properly line up his capsule for re-entry, and he complained: 'This devil ship! Nothing I lay my hands on works properly!' With its guidance systems off-line, the ship began to spin, and Komarov fired his attitude control jets, trying to regain control. Unfortunately the ship's designers had put the thrusters too close alongside the star tracker navigation sensors, and the delicate lenses were fogged and could no longer see the stars. Passing over the night side of earth, and searching for a more obvious reference target for his blurred instruments, Komarov had to use the moon in a desperate attempt to work out his alignment.

American National Security Agency (NSA) staff monitored Komarov's radio signals from a NATO facility near Istanbul. In August 1972, a former NSA analyst interviewed under the name 'Winslow Peck' (real name Perry Fellwock) gave a moving account of the interception. 'They knew they had problems for about two hours before Komarov died, and were fighting to correct them. We taped the dialogue. The Kremlin called Komarov personally. They were crying, and they told Komarov he was a hero. The guy's wife got on too, and they talked for a while. He told her how to handle their affairs, and what to do with the kids. It was pretty awful. Towards the last few minutes, he was falling apart … In a lot of ways, having the sort of job we did humanizes the Russians. You study them so much, and listen to them for so many hours that pretty soon you come to know them better than your own people.'

After Komarov's hectic re-entry, a small drogue parachute deployed out, but failed to pull the bigger 'chute from its storage bay. A back-up parachute was released, and became entangled with the first drogue, probably because the capsule was still spinning out of control. There was nothing to slow the descent, and Komarov slammed into the ground with all the force of an unrestrained 2.8-ton meteorite. He was killed instantly. NASA and its astronauts sent messages of sympathy, just as the Soviets had done for the Apollo 1 crew. Congressional snipers could no longer suggest that NASA had been uniquely or unusually careless in its preparations. Space exploration was a risk for both sides, and especially so in the febrile atmosphere of a race.

Keeping a grip

Webb decided he could no longer count on Seamans or Mueller to deal with North American. From now on he would personally handle NASA's relationships with the company. He and NASA's legal counsel Paul Dembling met in Washington on 26 April with Lee Atwood, and his counsel John Roscia, to discuss the crisis. It was an angry clash of chieftains. 'I'm not going to pay you any bonuses', Webb said. 'I can't face the American people and tell them we're going to pay the contractor bonuses after three people got burned up. You can sue me. You can do whatever you want. We're going to enter into a new contract.'

He was on the warpath. 'I thought that they had not dealt with the government in good faith ... They had a lot of people who didn't want to do things our way, but our way had to prevail if we were really going forward with Apollo. I told them they had to do certain things. They said they weren't going to do them. So I stood up at the table and I said, "All right, Lee, these are not negotiable. If you're not going to do them, I am going to take away every goddamn contract work that you have with us if I can, and give it to someone else." And I called in five other contractors and asked them to submit bids to finish Apollo.' A press release to that effect was issued at the end of April. It was an extraordinary gamble. If the lion's share of Apollo's construction had been transferred to some other company, it would have been extremely difficult for NASA to achieve the first moon landing before 1970.

Webb also made it clear that Harrison Storms had to step down as the chief of North American's space division. His old-fashioned gung-ho plane-builder's style of management had grated on NASA's clean-cut systems engineers and technocrats. The man who had won the largest federal contract ever granted to a single company was dropped from the team. As if by magic, Webb's threat to find an alternative contractor for Apollo began to evaporate (although he did call on the Boeing company to assume a wide-ranging supervisory role). Atwood had defended Storms as best he could, 'but he was not making it with the NASA people, and that was one of the real problems that came up after the Apollo fire. He was almost the scapegoat.'

Atwood was paddling furiously to salvage not just Apollo but his entire company. A merger was under way with another company, Rockwell, an automotive components manufacturer with ambitions to diversify into aerospace. What should have been a useful opportunity to recapitalise and expand became, instead, an opportunity for Rockwell to swallow North American wholesale, and on bargain-basement terms. Just as Webb had quite properly tried to warn Congress, the company's stock was suffering in the wake of publicity about the fire.

A question of integrity

On 9 May, Webb tried to begin that day's Congressional testimony with a pre-written speech detailing his reasons for still having failed to submit the Phillips report. He alluded to the increasing press interest and warned that if newspaper stories were generated every time some piece of trouble cropped up between NASA and its contractors: 'I do not believe we will have a successful space program. I think the ultimate result will be some kind of closure of the industrial system that is now very open with respect to things that are important, and on which it can be judged. Now, Mr Chairman –'

But Chairman Anderson felt moved to interrupt. 'Before you start with your prepared statement, I want you to realize this is not easy for the Committee either. You have to understand that people are tremendously interested in this whole situation and they ought to have the facts.'

Frustrated by leakages to the press from within NASA, Webb replied that when sensitive information is 'extracted from governmental or private industry through a process of sale or bribery to purloin documents, and then thrown into the public arena, this will destroy the system on which our success is built'.

Anderson agreed that some proper way of looking at the Phillips report was desirable, but by now, the whole question of formal submission was all but irrelevant. Most of it was already in the public domain, so Webb's interrogators quoted from it time and again, and then simply asked if those quotes 'as reported' did indeed appear in the actual document?

Webb's habitual ally Margaret Chase Smith was concerned about another matter. She wasn't on the attack. She was simply pursuing all the appropriate facts. 'I think you will recall that when you were here last, on April 17, I asked questions with respect to the selection of North American for the award of the Apollo space program … And your answer "yes" when I asked if North American was the first choice of the Source Evaluation Board is correct, is it not?'

Webb shifted uncomfortably. 'I thought you said Source Selection Board.'

'I'm sorry. Source Evaluation Board.'

Chase Smith was generous in allowing Webb a semantic hiding place, but the cat was out of the bag. He appeared to have been caught out in a direct lie. At the time of the original contract award in the autumn of 1962, Webb, Dryden and Seamans had justified their decision by referring to the excellent work North American had conducted on the X-15 rocket plane during NASA's first three years as a fledgling agency. These black dart-shaped machines, dropped from under the wing of a B-52 bomber, probed the very edges of space with incredible success and reliability, and North American had designed and manufactured them on time and on budget. Webb was concerned that the Martin company's proposal for Apollo was strong on academic theory, yet they lacked the experience of building rocket-propelled hardware with real live men aboard. The Triad's overturning of the Evaluation Board's choice had seemed nothing more than a perfectly allowable judgement call from the senior administrators of NASA who were ultimately responsible for the success or failure of such large-scale purchasing decisions.

Chase Smith's real question wouldn't go away. 'I wasn't questioning your authority to change the recommendation of the Source Evaluation Board. I was only trying to have the record set straight in this particular instance. I asked the question, and I thought you said that North American was second [in the scoring]. Was that right?'

Webb cited business propriety once again, and suggested a closed executive session in strict privacy, 'and then you can decide what questions you wish to ask in public'.

Smith was unimpressed. 'I think perhaps the Chairman and I may be aware of the answers, but I think the American public are the ones we are trying to get answers for.'

Chase Smith was interrogating him as fairly as she could in the circumstances. Webb knew perfectly well that her attack was not personal, nor motivated by self-advancement. 'One of the big contractors in the Apollo program had finally concluded they'd have to get rid of me, so they began to send their top dog around to see various senators. When they got to Margaret Chase Smith, she said, "Look, Mr Webb's an honest man and a good administrator. You'd better do exactly what he tells you, and nothing more."' But her questions stung all the more because of her honest reputation.

As the public hearings drew to a close, many people in Congress and beyond had gained, by now, an unfavourable impression of NASA and its chief. A popular quip doing the rounds was that the Agency's famous acronym stood for 'Never A Straight Answer'. New York Congressman John Wydler summed up the feelings of many politicians when he complained: 'You made statements or conclusions that may or may not have been right. I think it's our fault for taking you at face value, but we did. I think you have an obsession with secrecy, Mr Webb.'

The insults hit home. Lee Atwood met with Webb regularly during these weeks as NASA and North American desperately tried to rescue themselves and project Apollo from the chaos. 'He was a competent guy, an activist and a very brave man, but I never saw anyone quite so agitated as he became during those hearings.'

Seamans was also becoming concerned about his boss's state of mind. 'Jim took all this personally. He became terribly tense. Migraine headaches, which he tended to have anyway, were exacerbated. I had the feeling I was dealing with somebody who could explode at any moment.'

Webb's stress typified an agency-wide reaction. Everyone close to the Apollo programme was suffering. Mueller subsumed his grief in detail work, and in trying to 'hold together the psyches of the various people involved ... There are so many myriad ramifications that went through the organization during that time, and one could do a whole book on just the psychological reactions to the fire ... We don't do an adequate job of really recording the emotional turmoil that goes on.'

The astronaut squad above all felt the loss of their friends and colleagues. The man who best defended NASA's reputation at the hearings was Frank Borman, a Gemini veteran, a key member of the Accident Review Board, and a prime candidate for the lunar landing mission itself, if it ever came to pass. He testified several times before Congress, speaking in admirably plain language about the machinery, the risks, and above all, about his and his astronaut colleagues' intrinsic trust in the management of NASA, despite the fire. He stressed that in his experience the Agency had never knowingly 'overlooked, turned down or relegated to a lower priority for any reason whatsoever' any aspects of crew safety. 'In this case, unfortunately, we did not identify the hazards.' When it came to the underlying standards of the Apollo 1 spacecraft, for all its problems with communications and other balky hardware: 'The crew believed that everything that could have been done at that time had been done ... They were satisfied that they had a spacecraft that was not only adequate but safe for the test they were performing.' The fire, he explained, had been an accident, a terrible human failure but not the product of negligence. It was hard (and politically unwise) for Borman's questioners to be unmoved by his faith. If the astronauts felt that Apollo was an acceptable risk, then their opinion surely counted. NASA's critics on the Hill began to change the focus of their attacks from the spacecraft itself to the business deals that lay behind it.

Part Five
Dark Side of the Moon

Politicians may have been speaking the accusatory lines in Congress, but much of the script was spoon-fed to them by outsiders. Few lawmakers were particularly expert in the details of running the space agency, or the mind-boggling technicalities of the Apollo spacecraft itself. By contrast, some newspaper journalists had done a great deal of homework and were now setting the agenda. In particular, the *Washington Star* staff writer William Hines was adept at looking under NASA's hood and then feeding specific questions to receptive members of Congress, which they could fire off at the hearings. Apart from Mondale, Webb was frustrated by one other politician in particular, New York Congressman William Ryan, who 'figured the best way to get his name before the public was to criticize the space program. He sat there while Bill Hines fed him questions.' Hines's reporting was intelligent and could not be easily ignored. A major article in the Sunday edition of the *Star* on 21 May agitated Webb more than any other piece:

> *To persons familiar with NASA's organization, it is utterly inconceivable that Webb would not have known about such a far-reaching document as the Phillips Report.*

Referring to the savage covering letter in which Phillips expressed his unhappiness with Apollo's progress, Hines concluded:

> *It can be stated quite without question that this sort of letter simply would not be written, or such a report rendered to the agency's largest contractor, without the knowledge of the agency's administrator.*

Webb considered Hines a major irritant. On one occasion the reporter had a chance to challenge him face-to-face: 'Do you always tell the truth to Congress?'

'Mr Hines, I tell the truth better than you do in your reporting.'

Hines and other reporters also raised another question that was to haunt Webb as the hearings drew to a close in May 1967. Did a certain Fred B. Black Jr., a high-living chancer with a dubious reputation, play an unhealthy role in America's moon project?

Lee Atwood warily confessed to employing Black's services for a while. 'He was supposed to get all the information on how to advise people in Washington. We had hired him on some basis, some people's recommendation, I don't remember who. He was very well connected in Congress – or appeared to be – and we employed him for several years. He got us in trouble, and so eventually I had to let him go.'

Yes, Mr Black did get them in trouble. He got them all in trouble. Black's kind of trouble spread from a vending machine factory in Chicago that just happened to be co-owned by some of the decade's biggest crime bosses; it rippled through a bank in Oklahoma City that was effectively a cashbox for a powerful senator; it burrowed into the heart of the nation's space effort, and along the way it tainted Webb personally. It embroiled one of the most notoriously disreputable political figures of modern times, and finally – for this was the kind of trouble that knew no boundaries – it landed in the one place where, at all costs, it couldn't be allowed to spread its poison. Not if Jim Webb's NASA was to survive.

It landed on the President's desk.

The Black and Baker business

In March 2005, Washington insiders were saddened by the news that a popular socialite had died at the age of 80. Raised in Washington by a politically well-connected family, Nina Lunn Black first made her mark as a honey-blonde beauty in a short-lived movie career during the 1940s, but the actors strutting across Hollywood's plywood sets weren't half as fascinating to her as the political players on the Washington stage. She soon went back there, and became an indispensable fixture of the social scene. In 1948 she married the heir to a department-store fortune, then divorced him after a couple of years and wed a New York columnist at the opposite end of the financial spectrum. That union didn't last either. Nina's restless spirit kept her on the move through a series of liaisons that kept the gossip columnists busy for years. She was a beguiling, effervescent woman who enlivened Washington's often rather stuffy dinner-party circuit. She flirted and romanced with a number of men, but in the end it was her charming and roguish third and fourth husbands that she took the longest to walk away from, for they were one and the same fellow: Washington 'man-about-town' and lobbyist Fred B. Black Jr. After their first marriage in 1954, Nina and Fred kept a fine town house in the exclusive Spring Valley section of Washington, where people of influence could mingle in comfort and privacy. The backyard fence adjoined Lyndon Johnson's place.

The word 'lobbyist' neatly characterises an ill-defined political class of operator. They are not elected to any chamber, and therefore have no routine access to the rooms where power is supposedly wielded in Congress. In times past, they would quite literally accost representatives in the lobbies of the main chambers, trying to gain their attention on behalf of some cause or other. Lobbyists often show great passion and commitment, championing a wide range of issues that matter to ordinary people. But the professional lobbyist is a different matter. In the 20th century they became well-paid advocates for big business, and the sharpest of them work their magic in late-night clubs and bars and restaurants or in the executive suites of comfortable hotels. No corporation wishing to gain government contracts can hope to survive without them. However, the professional lobbyist's lack of genuine political commitment – the sense that they are little more than hired guns for whichever corporation happens to be paying them – has always tainted this occupation with a dark hint of the mercenary. Fred Black was surely one of the best-paid and most adventurous lobbyists ever to bang on NASA's door.

Harry Easley, a manager for Harry Truman's presidential campaign team, encountered Black in the 1940s, on the prowl for opportunities in Washington and exploiting the benefits of an earlier casual acquaintance with Truman. 'He was a very personable and fine-looking young fellow from Carterville, Missouri, and he played the ragged edges of things. He was a person who – well, his record was not entirely savory. He hadn't done anything particularly wrong outside of circulating hot checks and so forth, but he had unmitigated gall. He would tie himself to political people, and more often than not he would use them to his advantage for a job or whatever it might be.'

In the latter years of the war, Black made many friends in Congress and learned the ropes of defence contracting, but it was in the mid-1950s that he achieved his first notable coup. The accounts from hundreds of war contracts, large and small, were in limbo. Some companies insisted that even though the best part of a decade had gone by since they'd fulfilled their orders, the government still owed them money. More often it was the other way around. The government claimed it had been overcharged for war *matériel* by certain contractors, and now it wanted the money back. In 1956, the Howard Foundry of Chicago was faced with a $3 million 'excess profits' demand from the Justice Department. Faced with bankruptcy, Howard called on Black's services. He somehow persuaded a couple of Air Force procurement officers to take responsibility for collecting the debt on easier terms. Suddenly the whole thing was just an everyday accounting matter, and the Justice Department withdrew its dire warnings of prosecution. Eight years later, when Air Force officials were questioned about the affair (during one of Black's many courtroom appearances for tax evasion), they claimed to have recovered $800,000 plus interest from the Howard company. Beyond that, they offered no comment, although one official did let slip that 'we may be making Mr Black look pretty successful'. The rest of the debt was apparently 'lost in the shuffle'. Howard paid Black a fee of $150,000 and kept him on the payroll. After all, he had saved them the best part of $2 million.

North American Aviation's chief Dutch Kindelberger hired Black in 1959, officially to work under the company's public relations remit, and unofficially to see what influence he could bring to bear in Washington. Kindelberger's successor Lee Atwood kept him on during the Apollo bidding process. Atwood, a courtly, intelligent man, disliked Washington's underground networks, yet he was sanguine about the need for lobbyists. 'If you can know what projects are going forward for sure, it's an edge on the way you spend your money, the way you put your emphasis. It's an important commodity ... Ordinary information as to what a [military] service wants, what it needs and what it can pay for, should be a free commodity. It's in the best interests of everyone to know that. For a consultant to have to go to somebody who's friendly to find out something that other people can't find out – that may be corrupt. But – what is freedom of information? I don't know. But it's important.'

Somebody who's friendly ... Shortly after becoming Vice President and taking up the chair of the Space Council, Lyndon Johnson introduced Atwood to Robert 'Bobby' Baker, Secretary to the Senate Democrat Majority and Johnson's most trusted protégé. 'Bobby dropped in, carrying a dispatch or something. After he'd gone, Johnson said, "You know, that's a fine young man. If I had a son, I'd like him to be just like that." So there you are. He was really quite well connected.' Bobby Baker and Fred Black were close friends. The potential value of this alliance wasn't lost on Atwood and other senior executives at North American Aviation. But they had to be careful, for they were playing with fire ... Storms and his colleagues on the engineering floor despised Black. 'He would have sold his grandmother if he could make a buck.'

Baker knew Black as a 'superlobbyist' accustomed to ploughing his way through upwards of half a million dollars a year on a lavish lifestyle, seldom seeming to care where the next stash of cash might come from, and perfectly happy to gamble (and usually lose) thousands of dollars at a time on horses, or playing cards in the Las Vegas casinos. 'Half a million per year just wasn't enough money for Fred. He was a playboy of the first order. If he couldn't go first-class he wouldn't take the trip ... He had a quick eye and a grand way with shapely ladies. He loved booze. In short, he thought little more about tomorrow than did a fattening hog.'

That description might just as well have summed up the extraordinarily feline Bobby Baker himself. A poor boy from Pickens, South Carolina, he started his career in Washington as a Senate page in 1942, barely into his teens. In the 1950s, as Johnson ascended to the rank of Senate Majority Leader, Baker's fortunes rose in lockstep. 'Little Lyndon', he was sarcastically called. Johnson described him as 'the first man I see in the morning and the last man I see at night ... my strong right arm'. The bond between the two was remarkable, a surrogate version of a father–son relationship made yet more intense by their shared enthusiasm for political and financial intrigue. With Johnson now installed as Kennedy's Vice President, Baker was *the* man to see for anyone with a problem. In theory he retained his Senate job, answerable to the new Senate Majority Leader, Mike Mansfield. In practice he was still Lyndon's 'strong right arm'. He was also living a millionaire lifestyle, which was an odd circumstance for a young man supposedly earning just his modest Congressional salary of $19,600 a year. Somehow, he had acquired stakes in a law firm, a travel company, a Howard Johnson motel and dozens of other businesses. And in the summer of 1962, he welcomed half a dozen entire busloads of assorted celebrities and political friends (including Lyndon Johnson) to the opening of his proudest venture, his very own pleasure palace by the sea.

The Carousel Hotel in Ocean City, Maryland is a respectable place today, as innocent as apple pie – the perfect family destination, in fact. Its new owners, the Bethesda-based Hospitality Partners company, recently lavished $7 million on the 45-year-old building. It had been in a sorry state for some years. Armies of workers painted and redecorated inside and out, plumbing new bathrooms, laying fresh carpeting in all the public areas and dumping all the tired old beds and flickering televisions. A new seafront restaurant overlooks a cleaned-up beach, and a seventeen-foot-long replica of a pirate ship beckons kids to play innocent games. Tucked away in the back office, a few scrappy mementos are a reminder of racier times and a darker and more grown-up style of piracy. On the manager's desk, a fleshy cherubic face stares out from a framed *Life* magazine cover. 'Bobby Baker's Bombshell', the strapline reads. A wooden carnival horse mounted on the wall behind, flecked with peeling gold paint, is all that remains of the notorious Carousel nightclub where Washington insiders once partied until the early hours with a less than respectable crowd of casino types and assorted hucksters and party girls (there was an eight-strong team of scantily-clad cocktail waitresses on standby). If you were good to Bobby, he'd get you a good time. And in the morning you'd go down to breakfast and talk about what really counted: matters of business and money and power which were good for you, too.

On the surface all was well at the Carousel. Behind the scenes, Baker was struggling. The project was costing him $8,000 a month in loan repayments, especially after a savage Atlantic storm ruined much of the early building work, a circumstance for which he had been less than adequately insured. He turned for help to Senator Robert Kerr, another elder statesman (into his late sixties by now) who'd taken a liking to this eager, opportunistic young man. Before they got down to business, Kerr simply needed to know: how could Bobby Baker best serve the interests of Oklahoma? Baker decided it was time for Kerr to meet North American Aviation's main man in Washington, Fred Black. 'Kerr loved Baker. Bobby was his ears, nose, throat, hands and feet in the Senate. Kerr was dedicated to making Baker a millionaire, the same as he'd made Jim Webb a millionaire.' Well, maybe, except that so far as history can judge, Webb made his money within the law, while Baker's shaky fortunes depended mainly on his adventures outside it.

‘What’s in it for Oklahoma?’

Fred Black discovered that Senator Kerr treated certain banks as his personal toys. ‘He invited me to his office and we played some gin rummy. Then he asked me if I’d like some stock in a bank. He told me to write a check for $175,000. When I said I didn’t have it, he said, “I didn’t ask you that.”’ Kerr told Black to draw the check against the Fidelity National. Then he took out a special phone from the drawer of his desk and phoned through to the bank’s president. ‘He told him he had a new customer for the bank, and that the bank had just loaned me $180,000.’ The loan was to go towards Black’s purchase of stock in another bank, Farmers and Merchants in Tulsa, along with $5,000 in cash for walking-around money. There was no need for Black to offer anything tangible by way of security. ‘I’d borrowed the money, bought the stock, and then put up the stock to secure the original loan. It makes sense if you look at it long enough.’ Kerr’s sleight-of-hand was typical of his witty, mischievous style. Today it would be called ‘junk bonding’. Kerr’s largesse was motivated, of course, by wider concerns. According to Black, he had already discussed ‘if we, North American, wanted the Apollo program. Then he said, “If you get it, what can North American do for Oklahoma?” And I told him that I’d put plants in Oklahoma to build parts of Apollo or other type of work, so that Oklahoma would have less unemployment.’

North American’s sharpest salesman lost no time in approaching Webb. ‘Black was sent to me by Senator Kerr on three different occasions. We turned every one of them down.’ Nevertheless, as NASA moved inexorably towards awarding more and more of the Apollo hardware construction work to North American, Black’s reputation as a lobbyist was enhanced. Even Atwood admitted: ‘Senator Kerr worked on me to put more business in Oklahoma, and I think his implication was that he'd help. Fred used to have a nice house in Washington and he served dinner to me, and perhaps two or three other people. And Senator Kerr was there. I remember that very distinctly.'

In February 1962, the *Tulsa World* newspaper proudly announced that an ailing North American missile components factory was about to be rejuvenated for the space age. Its modest staff of 250 people was destined to expand to a thousand and possibly more. The centre-piece of their work was to be the Spacecraft Lunar Adapter (SLA, or 'slaw'). This conical shroud fitted between the top of the third and uppermost stage of the Saturn V rocket and the underneath of the Apollo command module. Inside the shroud, the delicate lunar landing module was safely protected against the stresses of launch. After reaching earth orbit, the upper stage of the Saturn booster fired its engine a second time to push Apollo towards the moon. Then the SLA fell away in four neat petals, and the lunar module was plucked out to begin its mission. The SLA was a relatively simple piece of kit, barely more than a protective shell, but it wasn't an insignificant one.

A year later, North American assigned construction of the outer fuselage of the 'service module' to its Tulsa plant. This stubby cylinder sat immediately behind the conical command module, feeding it with oxygen, water and electricity. It also housed the big bell-shaped engine that braked the capsule into lunar orbit, and at mission's end, pushed it away from the moon and towards home. Tulsa was also given the 'instrument adapter', a thin section of metal wall that fitted between the SLA and the Saturn's third stage. This housed the electronics and navigation systems (built elsewhere) to guide the booster's lift-off and ascent. Again, these adapters were fairly straightforward airframe shells rather than fully fitted-out machines, yet they were still important. Black appeared to have fulfilled at least some of his promises, capturing genuine slices of the Apollo cake for Kerr's home state. Black also now owned a modest piece of the very same bank, Farmers and Merchants, where the newly expanded Tulsa plant did its banking. Kerr-McGee's interest in the bank was much more significant. It held 60,000 of the Farmers' 160,000 shares.

Atwood had no special technical reason to favour Oklahoma for the SLA's fabrication. It was just that circumstances had conspired to make it an easy choice. 'Apollo was spread out over the whole country, with 400,000 people and so forth. Many companies put little feeder plants into other parts of the country, looking for good labor rates, a hospitable environment, uncrowded conditions, and finally, at least giving lip service to a social function – namely, spreading the work. How strong that is, I can't say. You can be as cynical as you like ... I did not promise Senator Kerr any work. He was almost like a salesman trying to convince me that we should work in Oklahoma. And he had a point, of course. He was a Senator and an influential one. This was a pitch and certainly we responded, if you want to look at it that way.'

It's difficult to judge, today, the extent to which Black's dealings may or may not have contributed to the main deal, the Apollo command module award to North American. As he later confessed, it wasn't necessarily his direct manipulation of the contract so much as his apparent '*aura* of influence' which worked in his favour. 'I never told anyone I could do anything that I couldn't do.' And he was always tactful dealing with people in public office. 'We don't want any favor', he would reassure them. 'We just want you to listen with an open mind.' It was all very well Jim Webb insisting that North American's track record on the X-15 rocket plane had won NASA's trust. 'I think it was a little more political. I think Mr Kerr asked for [the contract award] to be done that way.'

Webb was too much his own man to bend to anyone else's will. 'A lot of Senators and Congressmen talked to me about the Apollo contract. I was willing to talk to everyone about it. There are a lot of people who have tried to give the impression that I was not independent, and that I was in Senator Kerr's pocket.' Not so, he always insisted.

Kerr continued to support Apollo in his capacity as chairman of the Senate Space Committee until his sudden death in January 1963. His behind-the-scenes manipulations had been driven by instincts deeper than the mere love of profit, for he often delivered greater rewards to others than for himself, and he didn't always call back favours granted. Like so many politicians in his mould, according to Black: 'He always did more for other people than he asked other people to do for him. It was his philosophy, and naturally it put a lot of people in his debt.' Kerr's charm and outright shamelessness made him easy to like. He was the most appealing of the buccaneers surrounding Apollo, and genuinely supportive of its ambitions. Similarly, his vigorous championing in Congress of the COMSAT global communications satellite organisation was a fragrant mix of the visionary and the self-interested. As for the Apollo plant in Tulsa, how wrong could it be for a politician to make sure that his constituents back home played their rightful role in the space age?

By February 1963, almost exactly a month after Kerr's death, FBI investigators were on Black's trail. They were really interested in Bobby Baker's connections with him, rather than the NASA deals, but as a by-product of their investigations they taped Black on the phone to Dean McGee, the surviving partner in the Kerr-McGee company. Black was apologising because North American wasn't getting its new assembly plants into Oklahoma quite as fast as it had promised. He gave the impression that NASA's chief was somehow failing to deliver his side of the bargain. 'Since the old man [Kerr] died, this fellow Webb's gotten weaker and weaker where the state of Oklahoma is concerned. We sent him several things we wanted NASA to do. We got an okay on a third of what we wanted to put in there. He's just not going to do anything for us. NASA's not helping us. If the Senator was alive, he'd be helping … I want you to know, North American and Fred Black aren't backing up one inch.'

Mafia's moon

Gangsters of one kind or another seem to have cast their shadows over much of 1960s American politics. Millions of words, many of them hard to verify, have been written about the role supposedly played by the Mafia in the Kennedy era. The Mafia rigged his election. The Mafia (or was it the Teamsters?) got upset when Kennedy's brother Bobby broke implicit promises of gratitude and went after them in his capacity as Attorney General. The Mafia mourned the loss of their gambling interests in Cuba and collaborated with the US government against Castro. The Mafia, in league with the CIA, the FBI, and even LBJ, assassinated JFK. And so on. Why not boost the Mafia into space as well? Fred Black and Bobby Baker entered into various partnerships with several influential leisure industry entrepreneurs, including Mr Benjamin 'Benny' Sigelbaum, Mr Edward 'Eddie' Levinson and another Edward, 'Ed' Torres. 'All three men have gambling interests in Las Vegas', the *New York Times* reported, choosing its words with care. These three, in turn, were answerable to, or at least closely associated with, Meyer Lansky and Sam Giancana, two of the major players in America's covert economy at that time. These unorthodox businessmen never touched the space age directly, but their interests in the drinks and snacks vending business set a trap that would one day clamp around Jim Webb's ankles and draw blood.

The leisure industry loves one-armed bandits and slot machines because they sit in the corner of a room collecting money that no one necessarily counts too carefully afterwards. Vending machines were a popular business enterprise in the early 1960s, as the technique of automatically dispensing coffee and snacks became more mechanically sophisticated and self-contained, and the comparatively costly human staff required to service the machines became ever fewer. As

Bobby Baker scrabbled for cash to save his ailing Carousel Hotel, he and his friend Black were introduced to the advantages of vending machines. According to the accounts of both men, it was the inventive Senator Kerr who first came up with the idea. He told Black: 'I want to help Bobby. I'll get you the financing if you guys want to go into the vending business. There's a fortune to be made.' He arranged several loans from his favourite Fidelity National bank, amounting to some $400,000, to help Baker and Black on their way.

In October 1961, at the very time that the Apollo capsule contract awards were taking shape, North American found itself in the gunsights of a new vending company called Serv-U. Fred Black suggested that if Atwood allowed Serv-U's machines into his plants to feed and refresh his people, then various benefits would result. By early 1962, 40,000 North American workers, scattered among three large plants in California, were taking their coffee and snacks out of Serv-U vending machines. North American became Serv-U's largest customer, accounting for $2.5 million of its $3 million turnover, generated pretty much from scratch in less than two years. Atwood trusted Black's judgement, and he always claimed that North American remained entirely unaware that Serv-U was owned by Black and Baker, and their partners Sigelbaum and Levinson; or that the vending machines themselves were built at a plant in Chicago controlled by Giancana. Black recalled running much of Serv-U's business from a desk right alongside North American's publicity chief, Lee Taylor. 'He certainly knew about my connection.'

So far, so more or less normal in the way of American business, and no doubt business the world over. All these games might have remained a private matter had it not been for another vending company who wouldn't play ball. Capitol Vending was ousted entirely from a North American electronics sub-contractor, Melpar. And this was odd because Capitol had already paid Bobby Baker more than $5,600 in 'consultancy' fees to help get them into Melpar in the first place. He, in turn, had tried to buy out Capitol and absorb it into the new Serv-U empire, in what amounted to a hostile takeover. Capitol didn't want to sell. Nor did it take kindly to losing its 130 vending machine pitches inside Melpar's plant in Falls Church, Virginia. In a $300,000 damage suit, filed against Serv-U in September 1963,

Capitol claimed that Baker and Black had 'conspired maliciously' to force them out. Baker never imagined that a little court case over vending machines could end his career in Washington. The whole thing span wildly out of control during 1963, precipitating one of America's greatest political scandals and lending 'mankind's greatest adventure' an unsavoury touch of corruption.

Downfall

Baker's nemesis was as modest as his target was brash. Senator John Williams, a former chicken farmer and feed dealer from Millsboro, Delaware, spoke so softly on the Senate floor that reporters dubbed him 'Whispering Willie'. He had never attended college, yet when he talked people listened. He exposed some of the biggest corruption scandals of the late 1940s, 50s and 60s, sending dozens of government officials to prison and saving taxpayers untold millions of dollars. He had no special staff and paid all his own expenses. This was a man obsessed with personal probity. And he was no mere political point-scorer. The subjects of his investigations squirmed on both sides of the bi-partisan fence. 'A man that's going to be crooked is not going to be crooked just because he's a Republican or a Democrat', he once said. Williams was also courteous to his victims, informing them in person about his discoveries and giving them a chance to repair some of the damage, or at least resign with dignity, if they didn't want to face public humiliation. In nearly a quarter of a century of sleuthing, he never once laid a false accusation. Modest to a fault, he had no interest in bragging about his victories. 'I always figured that if there was something newsworthy, the press people would find it, and if they don't find it, it isn't newsworthy. I didn't need public relations.' On one notably ironic occasion, the IRS accused Williams of defaulting on some of his taxes. He not only vindicated himself but discovered that the IRS official who had laid the charge had embezzled $30,000 by juggling taxpayer accounts. He wasn't the only one. Eventually, 125 IRS staffers went to prison and nearly 400 resigned.

In September 1963, Williams demanded that the Senate Rules Committee investigate Baker's role in the Capitol Vending case. He encountered some reluctance to pursue the matter. Eventually he won a compromise: an investigation could proceed so long as it applied only to Senate staffers and not senators themselves. 'SUIT AGAINST AIDE DISTURBS SENATE', the *New York Times* reported. 'Senior Democrats have been made uncomfortable by charges of influence peddling that have been lodged against one of their most popular employees.' There was little need for observers to read between the lines.

By October, Williams' investigations had forced Baker's hand. He resigned his post as Secretary to the Senate Democrat Majority. Over the next four years he was hounded by Congress, along with the FBI and the IRS. Baker's plea to Bobby Kennedy not to let him go down in case he revealed too much about John F. Kennedy's risky love affairs, which he had occasionally engineered, was unsuccessful. The onrush of accusations against him was too momentous for anyone to halt – and besides, the Kennedy clan was watching with a less than sympathetic gaze to see if the scandal might damage Johnson and remove him as Jack Kennedy's running mate in the 1964 election.

On 22 November, Baker's pathetic mess was eclipsed by an immensely greater disaster: President Kennedy's assassination. While the entire country agonised and grieved, North Carolina Senator Everett Jordan, chairing the Senate Rules Committee, was forced by a relentless John Williams to keep his focus on Baker. In a forlorn call to Johnson just two weeks after he had been sworn in as President in the wake of Kennedy's assassination, Jordan promised that he really was 'trying to keep the Bobby thing from spreading ... it might spread a place where we don't want it to spread. It's mighty hard to put out a fire when it gets out of control.' Baker was unable to call upon support from his old father figure and mentor, who by now was desperate to keep as far away as possible from the scandal in case it destroyed his fragile inheritance of the presidency.

And the fire was spreading fast. As a result of the remorseless digging into Baker's affairs, the 'Savings and Loan' scandal emerged as the biggest shock to Washington in those far-off days before Watergate. In July 1962, there had been moves by the Kennedy administration and by Congress to impose tax increases on the savings and loan industry. That same summer, Baker had gone on a trawl of S & L company bosses, soliciting campaign funds for senators and members of Congress, Democrat and Republican alike, who might be sympathetic to their cause; not least, Senator Kerr, who'd promoted the tax legislation in the first place, but who apparently might be enjoined to change his mind with a little encouragement. Baker implied that a modest cash investment would ease the S & L people's burdens. He pocketed large sums of the collected cash for himself, amounting to $80,000 and probably more. He then used his ad hoc dollars to prop up the beleaguered Carousel Hotel project. In September 1962, the terms of the S & L tax bill were substantially softened as it moved through Congress, once again with Kerr at the helm. The following month, Baker collected another $50,000 from grateful campaign contributors. At least $16,000 more followed in November. Then another $33,000.

Much of this great scandal emerged purely as a consequence of Capitol Vending's relatively minor lawsuit against Serv-U. They soon withdrew their case after Baker settled with them for $30,000, (a mere one-tenth of their original demand), yet the can of worms that the affair had opened simply wouldn't seal up again once the lid was off. Bob Kerr's name featured repeatedly in the unfolding drama, but he was dead, and therefore unable to verify or deny Baker's claim that he had properly delivered the S & L 'campaign funds' to Kerr, who had then loaned him cash 'which he promised to replenish out of his personal funds'. No one could check Baker's heartbreaking memories of Kerr in the closing weeks of 1962, ill and close to death, yet fondly concerned for his protégé. 'He wanted to let me know he loved me and loved my family. He said, "Bob, I hope this is going to be the best Christmas ever for you. I know you've had it tough." And he said he wanted to sort of wipe the slate clean. "Just put what you owe me down to legal expenses, or a fee for all you've done for me."' No one was convinced, although it might possibly have been true.

Throughout 1964, Lyndon Johnson was acutely aware that his first year in the White House had been inherited, not earned. He was looking ahead to re-election in his own right. This was a sensitive time for him, and the Baker business was profoundly unwelcome. He telephoned Robert Byrd (no relation to Harry Byrd), a tame senator he'd finessed onto the Rules Committee some while before. 'Now listen, Bob. D'you think I'm in good shape in your committee, your Rules Committee?'

'Aw, darned if I know. I think so.'

'Well, if you don't know, what the hell are you doing up there on that committee? I put you *on* that so you *would* know!'

Rules Committee chairman Everett Jordan was more direct, warning Johnson: 'They want to get into campaign expenses, Baker putting out money to senators, controlling what's going on in committees – it's the damnedest mess you ever saw. The press just eats it up ... I need some help, and I need it bad. Maybe if you called Mike –'

Johnson phoned Mike Mansfield, the Senate Majority Leader who'd been forced to sack Baker. 'They're going to make all the political capital they can out of this', Johnson complained. 'And we'll just have to go after [Senator John] Williams. He's a mean, vicious man. I saw him on television. He's emotionally unbalanced and he's so goddamned interested ... I don't know why he's so worked up about people's income. But sometimes these holier-than-thou ones –' Johnson let the sentence trail off. He was too indignant to put his vitriol into words. There was no longer anything he or anyone else could do, except let the drama play itself out.

Black's back taxes

Meanwhile, the questionable legality of the FBI's surveillance kept Black on the streets, for the time being at least. The wiretappers had never primarily been interested in him, but the FBI decided it might as well alert the tax authorities to the possibility that he had accumulated a substantial undeclared income from his associations with Baker. From 1963 and the breaking of the Capitol Vending story, to the end of the decade and the final closure of the epic Congressional investigation into Bobby Baker's affairs, Fred was in and out of court a dozen times. In June 1964 he sued North American Aviation for $9 million in damages, claiming he'd been fired unreasonably. A month later, North American sued him back, claiming he'd violated his contracts with them by failing to declare his other business interests. One way and another, the desired $9 million compensation wasn't forthcoming.

That same year, Black was tried for tax evasion. The District Court judge in this case was John Sirica (destined a decade later to preside over the trial of the Watergate burglars). He was moderately sympathetic to Black, who claimed to have spent a great deal on non-taxable expenses rather than personal income. Sirica suggested that big businesses such as North American who'd employed his skills had used him as 'the scapegoat to cover their own slimy trails. And if that is so – if they hired a man to do their dirty work – they should be here in the courtroom with him ... They're not fooling anybody. I know how some of these companies work in Washington. Shouldn't they have kept some record of the money? You'd think they might have known they would be asked to account for these sums. Maybe Mr Black gave this money to someone but he can't account for it?' In the end, the jury remained unconvinced and found Black guilty of evading $91,000 in taxes. Sirica sentenced him to fifteen months in jail. Smartly turned out in a blue suit and white tie, Black remained unruffled throughout. When asked if he had any last comments before sentencing, he simply said: 'As I told you before, I'm innocent.'

The trial caused yet more unease in the White House, especially when Black's early dealings with the Howard Foundry of Chicago emerged. On 21 April, after reading some unfavourable opinion polls about the moral tenor of his administration, President Johnson had put in a tetchy call to Robert McNamara's new deputy Cyrus Vance:

'There's some testimony in the Black trial today. Are you familiar with it?'

'No, sir, I haven't seen it yet.'

'It says that some company paid Fred Black some money to bribe [Air Force] officers ... I just wanted to be sure that if there's any bribery going on, we know about it. We've got to watch that. [Did you] notice the Harris poll? Half of them think I can't keep corruption out of the government ... So I don't want one case of corruption. I don't want anything close to it. That means my mother and my brother and my dead daddy, and everybody else ... If it looks suspicious, kick it out.'

Black's subsequent appeal in 1965 eventually reached all the way to the Supreme Court, for it cut to the heart of America's internal surveillance culture. Solicitor General Thurgood Marshall admitted in a written memo to the Court that evidence obtained from a wiretap directed at another target was probably inadmissible in Black's case. Yet another year dragged by, and the Court eventually quashed Black's conviction. In 1969, the Justice Department lost patience and filed another suit against him. Black had by now admitted to at least some of his tax liability from the North American and Serv-U years, amounting to $140,000, but so far, half a decade later, he'd paid only $676 and 30 cent's worth. Once more he stepped out of the quagmire a free man. In fact he stayed out of jail for the best part of the next three decades.

In the normal way of things, the Great American Society could always accommodate a happy-go-lucky chancer like him. A few high-profile tax fraud cases never did anyone's reputation any harm, so long as the amounts were sufficiently respectable; and in retrospect, even his louche contributions to America's space effort didn't seem entirely bad. After all, North American did eventually build a fine capsule, and none of the deals he'd pulled off had contributed directly to the fire in the prototype version – which turned out, remember, to have been more NASA's fault than anyone else's. He remained unrepentant. 'North American was five times as large as Martin [the other company bidding for Apollo in 1962]. When they got the contract, you had five times as many happy people as you had unhappy people. That's democracy in its finest form.'

Bad timing

The final verdict against Bobby Baker was delivered in US District Courtroom 21 on the morning of 29 January 1967. The various investigations against him had spanned more than three years, but this jury needed to deliberate for only seven hours. He was found guilty of eight out of nine charges against him, ranging from tax evasion to fraud and larceny.

Television and newspaper reporters eagerly reminded their audiences how the Serv-U business had precipitated Baker's five-year plunge from grace. From NASA's point of view, the timing of the final verdict against him couldn't possibly have been more unfortunate. The Apollo capsule had exploded just two days earlier, and every last aspect of North American Aviation's history was now under the spotlight, including the Serv-U business. Webb wanted to clear up any misunderstandings before the rumours span out of control. If he allowed the stories about Fred Black to continue circulating before he'd had a chance to make a formal response to the Senate, he knew the consequences could be appalling. At his urgent request, Clinton Anderson held a closed executive session of the Senate Space Committee on 12 June. Republican Senator Edward Brooke of Massachusetts (the first African American ever to be elected to the Senate) opened the questioning of Webb:

'Now, I have to ask you these questions', he began. 'Were you ever a stockholder in the Serve-U Vending Company?'

'No, sir.'

'Were you ever a director?'

'No, sir.'

'Did you negotiate the contracts by which the vending machines were put into North American?'

'No, sir. I didn't know about it until I read about it in the papers.'

When asked if Black had tried to influence Apollo, and whether or not politics had entered into his decision-making, Webb admitted frankly that yes, of course Black had been to see him, and of course his authorisation of money and resources was a matter of great concern to politicians. Yet he insisted, as always, that the final awards of all construction contracts were based on merit alone. At the time, North American had seemed the best candidate to build the capsule.

Webb's conscience was clear. 'There was no problem as far as we were concerned. The problem was with people who were looking for something to burn Mr Johnson or Senator Kerr [his posthumous reputation] or the Democrats. They were determined that there was something wrong, and they were going to find it.' But many politicians, Democrats and Republicans alike, became suspicious of Webb himself. It was an incontrovertible fact, for instance, that during his time with Kerr-McGee, Webb was 'asked to go on the board of the Fidelity Bank and Trust Company, which was a small independent bank. I told them I couldn't afford to do that just for director's fees. Unless [I had] some business interest, I couldn't do it. That worked forward to the point that I and other people in Kerr-McGee bought a substantial interest in the bank, and I became a director representing our group.' So far, no problem. But when Webb assumed command of NASA in 1961, he still retained $600,000's worth of stock in Fidelity National – the same bank that had financed Black and Baker's Serv-U operations in such an unusual manner, and which also won the account of North American's plant in Tulsa. 'Now, when I became Space Administrator, I did not sell the Fidelity stock. The last thing that ever entered my mind was that a bank in Oklahoma City would be involved in any way with the space business. It turned out that Senator Kerr arranged for that very bank to loan money to Bobby Baker in a substantial amount. So everyone tried to draw the association that I owned stock in the bank, Baker got a loan from the bank, and the vending machines went into North American. So it was a question of association, but there was nothing to it.'

Denied their big kill, a couple of senators resorted to nit-picking. They wondered about the 'per diems' routinely claimed by a business executive or government official when travelling away from home.

These little perks didn't usually amount to much, unless they were abused. Some people liked to buy Christmas presents, bottles of booze or even call girls, and then claim the money back in per diems disguised as innocent expenditures on food and hospitality.

'Did you stay in a hotel in Tulsa when you made a speech out there?' Webb was asked.

'Yes sir, I stayed in the Chamber of Commerce suite.'

'Did you collect your per diems?'

'No, sir.'

'How do you know?'

'I never collect my per diems.'

No one scored any direct hits against Webb. Nevertheless, the Senate Space Committee's final report into the Apollo fire had sharp words to say about his honesty and propriety, even if, for form's sake, the text referred to NASA as a whole rather than directly to its chief:

> *NASA's curious reticence to supply facts and materials relevant to a thorough evaluation of Apollo program management brought the credibility of NASA and its top management into sharp question. This lack of candor threatened one of the essential assets of the space program – the confidence of the American public and their elected representatives. It will be difficult to restore the trust forfeited.*

Mondale insisted on his own three-page addendum to the report, in which he said outright that: 'Officials attempted to mislead the Committee and evaded giving frank answers'. He scorned NASA's 'solicitous concern for corporate sensitivities at a time of national tragedy' and berated its 'refusal to respond fully and forthrightly to legitimate inquiries'. He meant Webb, of course.

The wrong kind of business

Black's sorriest mistake, in his later years, was to get involved with the Colombian drugs business. This was a trade which could not be forgiven, for even in Washington there could be no disguising cocaine as a source of valuable manufacturing contracts, a bringer of jobs to a grateful politician's home state, or in any other way a benefit for the common wealth. Nevertheless, picture an indefatigably cheerful and neatly turned-out charmer in his mid-sixties, slipping out of his apartment in the Watergate complex every few days with an attaché case containing a good few tens of thousands of dollars in cash. Then he spends a relaxed afternoon hanging around a convenient bank, depositing innocent-looking modest sums into a couple of dummy accounts. He doesn't have to walk far. There's a branch of Riggs National situated right there in the Watergate. It also helps that the bank's vice president is in on the deal. Over a period of five years, between 1977 and 1982, Black laundered at least $1.5 million in Colombian drugs cash, but on 8 July 1985 his luck finally deserted him. Nina had divorced him (again), her patience finally exhausted; his rent for the Watergate apartment was four months overdue, and worst of all, a tireless team of prosecutors bearing down on the drugs gang had unearthed a convincing star witness: none other than the leader of the gang, plea-bargaining a lighter sentence for himself at Black's expense. Turning to one of his unfortunate co-conspirators as he prepared for the last of his many court appearances – this time he wouldn't be coming out – Black said: 'Well, I'm sorry we got clobbered. Keep your chin up.'

In the courtroom, US District Judge Thomas Hogan described Black as a 'sophisticated, intelligent man-about-Washington' whose long-overdue account now had to be paid. He was 'morally bankrupt, a perpetual criminal … Whether through greed or to make your one big score in life, you've led a very fine life for at least ten years with no visible means of support.' In between his bank runs, Black had stuffed fat bundles of cash into his kitchen freezer for overnight safekeeping. 'It's beyond imagination that you didn't realize that some part of these funds came from narcotics. Businessmen simply don't do business this way', said the judge.

Black served seven years in jail for drug running, and for evading close on half a million dollars in taxes. The IRS had at last won its revenge. He died in January 1993, a few years after his release, at a nursing home in Maryland, aged 80. Love him or hate him, the earthly world of space politics had lost one of its most adventurous characters. Perhaps the kindest thing that can be said about him was that, come rain or shine, he took life on the chin and never showed any self-pity. The worst one could say is that the space business never seemed to touch him emotionally or philosophically. It might just as well have been an extremely large drugs operation.

Part Six
Recovery

Quite apart from dealing with the consequences of the fire, Webb also had to defend some of the unmanned planetary missions whose regularity we take so much for granted today. In the mid-1960s, the technology to touch down on Mars or photograph the swirling atmospheres of Jupiter, Saturn and other planets was still in its infancy. Webb had defended the Jet Propulsion Laboratory's unmanned projects as best he could, up until the Apollo fire. On 26 July 1967, the Senate Space Committee grilled him about cutting back on NASA's plans for the Apollo–Saturn hardware once the lunar missions were completed, a scheme known as 'Apollo Applications'. He was asked to choose between that programme and an unmanned probe called Voyager, designed to be lifted two at a time aboard a Saturn rocket and sent to Mars. 'Apollo Applications is a small investment to expand on something you've already spent $15 billion to get, and it seems to me this is important', he challenged; and on the other hand, if the United States retired from the robotic exploration of the planetary field, then the Russians would carry on. 'I'm not going to give aid and comfort to anyone to cut a program. I think it's essential we do both.' Congress thought otherwise, and the Voyager was cancelled. However, it was reanimated in later years as a 'grand tour' of the solar system's outer planets, using a smaller rocket. Likewise, the Mars missions went ahead in a slightly different guise (the Viking project) and under JPL's direction. We have to remember that Webb had very few examples of interplanetary mission success to back him up in those days, and certainly none of the stunning close-up colour views of alien landscapes so familiar to us now.

How, then, to maintain the momentum of Apollo itself now that the fire had dented NASA's credibility, and space was slipping even further down the agenda of Johnson's troubled presidency? Webb now made one of his less courageous decisions. He decided that the Apollo Applications Program had to be scaled down internally by NASA itself, before Congress could do any more damage from the outside. It was a bitter choice. In the time of innocence before the fire, 'Seamans and I felt very strongly that, having perfected the Saturn–Apollo, we should use it for a limited amount of time to gain operating experience, to see what could be done in working with very heavy payloads and long distances'. These plans, greatly desired by many people within NASA, centred on a large multi-purpose space station and allowed for retaining Saturn as a possible booster for future manned Mars missions. How could NASA aim for the stars while it had yet to draw down the moon? Congress was no longer willing to be seduced by such dreams.

By 1967, NASA had contracted for a total of fifteen Saturn V boosters. Webb believed that, even if this limited production run was to be completed, he would have to persuade Congress that every last one of them was necessary to ensure Apollo's success. And maybe the fire had dented his own confidence in NASA. 'I didn't want the burden of seeking approval for a major follow-on project when we might be faced with one or two unsuccessful attempts at making the landing.' If NASA started making plans to use some of the rockets for other purposes, it might appear that Apollo as a specifically lunar project was running at excess capacity and should be cut back. Allies had warned Webb of the changing mood in Congress. Even the powerful Texan who had so happily secured the Manned Spacecraft Center for his backyard in 1962 felt he had to issue a caution in these less extravagant times. 'Al Thomas told me there were a number of people in Congress who would not support NASA in the things we needed to finish Apollo, if they felt that we were trying to get a foot in the door for a large follow-on project.' Webb listened sympathetically to plans for new missions, but his travails in Congress had dented his ability to share the space cadets' hopes.

The harassed Jim Webb of 1967 was certainly not the same man who'd pledged, back in 1961, that he would tolerate no interference in NASA's technical decisions. He was now watching his back while bending to the political wind: uncomfortable contortions for someone so used to commanding events rather than being commanded by them. The man who had once argued with Kennedy that the moon should be only one target in an overall quest for preeminence in space now prioritised the moon above all else. He hated making this judgement call, and persuaded himself that the curtailment of the Saturn production run might only be temporary. He was desperately sympathetic, too, towards von Braun. 'Wernher was very disappointed because he couldn't see anything coming along after [the lunar landings]. Therefore he knew he was going to have to start disbanding his outfit.' Not that Webb was ever concerned with preserving jobs just for the sake of it. Although one of his principal ambitions with NASA was to spur the economy and provide a source of employment, those jobs still had to be justified by the tasks at hand. There could be no 'make-work' in Webb's world.

NASA's second-in-command Bob Seamans certainly felt a personal loss of 'impetus'. His relationship with Webb hadn't recovered, and he found himself being excluded from significant meetings, and in other ways simply no longer a full partner in the leadership. After the fire, Webb had called on NASA's nuclear propulsion chief, Harold Finger, to create a new Office of Organization and Management at headquarters, charged with supervising the other programme offices and making sure that procedures were properly followed. Supposedly answerable to Seamans, Finger was, in fact, Webb's new right-hand man. On 2 October 1967, Seamans decided he'd had enough of this insult to his authority. He personally handed Webb his letter of resignation.

Webb suddenly asked him: 'What do you think your peers will say about the job you've done over the years here at NASA?'

Caught off-guard, Seamans could only reply: 'I think they'll feel that I did a satisfactory job.'

At that moment Webb's parting shot seemed a little insulting, a veiled accusation of failure. Eventually Seamans would understand that the question had been genuinely meant. Legacy was important to Webb, and he supposed that it must be to his colleagues as well. In future years, the two men would regain their close friendship. But not for now.

Webb went immediately over to the White House and informed

President Johnson of Seamans' departure. NASA was suddenly in need of a new deputy at a time when – once again – no one of any calibre wanted that kind of job. Webb called around a few people he knew and had a couple of straight turndowns. He and Johnson soon agreed on Thomas O. Paine, a materials engineer from General Electric, and a man who was keen to broaden his experience with a government post. Characteristically, when Johnson asked Webb, 'Is this the man you want?', Webb replied: 'I want him, but he has to be your appointee.' Paine would prove entirely satisfactory to Webb. He was fast on the uptake at NASA (he had to be), trustworthy in reporting problems and careful in his bureaucratic dealings with Congress. At the same time he was unafraid to shoulder bold decisions. Just as well, because he would soon be called on to authorise one of the greatest gambles NASA has ever taken.

Around the moon

In the words of space historian Roger Launius, Webb had 'politicked, coaxed, cajoled, and maneuvered for NASA in Washington' to the best of his ability. He had pulled NASA from out of the fire, but his political credit was exhausted. In order to replenish his influence, Apollo had to be reanimated from its ashes. There could be no more brave words, no more subtle arguments or debates about the engineering. Something actual had to lift off the pad and fly – and soon.

North American's command module had proved the most seriously flawed element in the Apollo system, but it was by no means the only problematic piece of hardware delivered to NASA. In June 1967, the first lunar module from the Grumman company in Bethpage, Long Island arrived at the Kennedy launch centre to be mated with a Saturn booster for a test of its systems. Would it fit properly inside its SLA shroud, communicate with the rest of the rocket and basically prove itself ready for a mission? Well, no. The lunar module's fuel tanks leaked. The machine designed actually to land men on the moon wasn't fit to fly. Rocco Petrone, NASA's blunt-talking chief of launch operations, tore into Grumman's representatives. 'What kind of two-bit garbage are you running up in Bethpage? What kind of so-called "tests" do they do before sending this wreck to us? They'd better get this fixed, and fast! Your name is mud around here until they do.'

There was one last ray of hope. By November 1967, the gigantic Saturn V rocket was ready for its first test, albeit with only an unmanned 'boilerplate' version of the troubled Apollo capsule at its tip; yet when technicians at the Florida launch complex had stacked all three stages of the 3,000-tonne booster to its full height of 364 feet for the first time, nobody really believed in their hearts that such a monster could ever leave the ground. As if to confirm their worst suspicions, the six-day countdown stretched to seventeen as endless failures cropped up on the launch controllers' screens. Time and again, Petrone tried to get the countdown running smoothly. The fuel pumps failed. Batteries on the second stage's control system failed. The computers crashed. It seemed impossible to get all the thousands of systems working in harmony. Perhaps Saturn V was just too huge and complicated ever to work? An exhausted Petrone muttered: 'Can we ever get this baby to go? Can we ever get all the green lights to come on at the same time?'

At last, in the early morning of 9 November, the world's largest and most powerful rocket lifted off successfully, in a mission designated Apollo 4. NASA press officers announced that Apollo was back on track, at least as far as the launch vehicle was concerned. Unfortunately, the Saturn still had some nasty surprises up its sleeve.

On 4 April 1968, a second Saturn V took off on a test flight. From the press stands five miles distant from the pad, everything looked wonderful. The five thunderous F-1 engines lit up and the Saturn rose majestically into the sky on a gigantic column of fire. NASA's press officers announced another brilliant success. As it turned out, few newspapers or TV stations showed much interest in the launch because an earthly tragedy had turned their attention elsewhere. Civil rights leader Martin Luther King was shot and killed that same day. If the press hadn't been so distracted by King's assassination, they might have asked a few questions about this second Saturn V flight. Two of the five F-1 main engines had failed shortly after lift-off, leaving the rocket dangerously unstable. Mission controllers kept their fingers over their 'abort' buttons, because at one point (too high in the trajectory for civilian observers to notice) the rocket keeled over and headed nose-down towards the ground before levelling out again. Somehow its guidance computers compensated for the failed engines, and the upper stage, with its unmanned Apollo capsule, just about staggered into orbit.

It wasn't exactly dignified. After careening like a drunken elephant, the Saturn's upper stage ended up flying backwards around the earth. The dummy Apollo capsule sustained heavy vibrations during the launch, and several of its monitoring instruments were shaken into radio silence. Meanwhile, the conical SLA section beneath the capsule, designed to house the spider-like lunar landing craft in future missions, had lost one of its four 'petals', leaving the Apollo and upper stage combination dangerously weakened. The final phase of the mission was to relight the Saturn upper stage's engine by remote control and see if, in theory, it could boost the dummy Apollo in the moon's direction. The engine wouldn't even start.

The flight was essentially a failure. If there had been astronauts aboard, they would have been severely endangered. At best, they would have had to fire the small escape rockets attached to the nose of their capsule and make a dangerous bid for an emergency splash-down. The worst didn't bear thinking about. Saturn V apparently was no more ready to fly than the capsule or the lunar lander.

Thousands of engineers were immediately put on a 'crash programme' to fix the Saturn. Analysing the failures was tricky because there was no wreckage from the last launch to examine (unless one counted the expended first and second stages lying at the bottom of the Atlantic). The only evidence was radio telemetry data describing the flight in numbers and symbols on long rolls of computer printout. From this slender evidence, an incredible detective effort resolved the problem. Kerosene and oxygen fuel thrumming at high speed through the five main engines' pipework had set up a resonance that shattered the fuel lines. Think of the effect in a tiled bathroom when someone whistles at a certain pitch and the whole room suddenly hums. Make the tiles out of several different materials, and the humming can be eliminated. With a similar theory in mind, the Marshall Center engineers and their contractors at Rocketdyne redesigned the F-1 engines with a new mix of alloys, and the vibrations in the fuel lines disappeared. A small percentage of inert helium was also pumped into the fuel mix to dampen the frequencies as the fuel swooshed through the pipes.

It was a brilliant effort: probably one of the cleverest and fastest (and least-known) engineering achievements in history. Barely a month after the second Saturn's shuddering flight, ground tests of the redesigned engines confirmed that the fix had worked. George Mueller's 'all-up' testing concept had been vindicated. All the engineering data required to identify the Saturn's faults had indeed been made available from the radio telemetry, and there had been no need to launch the stages one by one on separate test flights. Mueller's position in the Apollo hierarchy was safe, although Webb continued to be distrustful of him on a personal basis. 'Even if I wanted to, I couldn't fire him because he was manager of our successful fast-moving Apollo project, and one of the ablest men in the world ... The last thing I wanted to do was to lose him, but I also had another desire, which was not to let his way of working make too many difficulties.'

In the light of this recovery effort, George Low, the man Webb had appointed as Joe Shea's replacement, had an incredible idea. His plan was this. Once the improved version of the Apollo capsule had been verified in earth orbit, its very next mission should take it all the way to the moon aboard a Saturn V.

Low's reasoning was simple. Apollo's astronauts and mission controllers might as well gain experience in deep space navigation and lunar orbit procedures, even if the lunar module wasn't yet ready to fly. The CIA and other US intelligence agencies knew that the Russians were belatedly attempting a lunar landing of their own. While their troubled N-1 lunar landing booster was no great threat to the Apollo timetable, a smaller Russian rocket presented more of a challenge. The Proton, still in use today, was at an advanced level of development by 1968. More significantly, the Soyuz was nearly ready to enter routine service. Many lessons had been learned from Komarov's death, and the capsule was set for a long (and largely reliable) career as a 'workhorse'.

If only Sergei Korolev, the great 'Chief Designer' of Soviet rockets were still alive, Russia's bid for the moon might have succeeded, but he died in January 1966, the victim of a botched hospital operation. The brilliant man who had sent Sputnik and Yuri Gagarin into orbit was gone, and the Soviet space effort was a shambles, with 26 separate industrial and governmental agencies squabbling to control Korolev's legacy. Nevertheless, the fatal problems with Soyuz that had killed Vladimir Komarov proved relatively easy to fix. There was a sporting chance that the new Proton rocket, designed by Korolev's rival Vladimir Chelomei, could send a Soyuz around the moon by the end of 1968, even if it had to be so crammed full of extra fuel and oxygen that perhaps only one cosmonaut could fit inside. If this mission succeeded, it would be yet another Communist triumph: first satellite in space, first animal, first man, first space-walker, first woman, and now, the prospect of sending the first man to the moon. No matter that the Russians had no real chance of making a touchdown. The mere presence of yet another Soviet hero in the moon's general vicinity would clinch the space race in the public's imagination.

In the second week of August 1968, George Low held a discreet meeting with Manned Spacecraft Center director Bob Gilruth and mission control supremo Chris Kraft. It took them about ten seconds to see the merit of Low's plan. Likewise, astronaut Frank Borman and his crewmates Jim Lovell and Bill Anders, originally scheduled to make the second earth orbital test of Apollo, lost no time in volunteering to take their ship all the way to the moon instead. It certainly was a risky proposition, but no one could bear the idea of losing out yet again to the Russians. In March that year, Russia had launched an unmanned Proton rocket deep into space, carrying a so-called 'Zond' spacecraft in the opposite direction from the moon, yet on a trajectory clearly designed to test a Soyuz capsule's re-entry from such a distance. No one was quite sure if its strange trajectory was caused by a failure in the navigation instructions, or constituted a deliberate ploy to confuse American watchers. Even so, it was enough to scare NASA. Who knew what the Russians might come up with next?

On 15 August, George Mueller, Sam Phillips and NASA's new deputy Tom Paine telephoned Webb in Vienna, where he was attending an astronautical conference. They had considered sending Phillips to Vienna in person, but hadn't wanted to generate ripples of curiosity among any Soviet delegates at the conference. Webb first got to hear about Low's scheme in a conference call to a secure line in the US Embassy in Vienna. His initial reaction was one of outright fury. 'Are you out of your mind? You're putting our agency and the whole Apollo project at risk!'

Phillips recalled: 'If a person's shock could be transmitted over the telephone, I'd probably have been shot in the head.'

When Webb took time to think about the scheme, and to consult with his informants at the CIA, he began to see merit in it. Yet he also felt that this momentous decision should not be his alone to make. After seven years in harness, it was time for him to think about stepping down from NASA.

Departure

Webb had been brooding for some weeks about how to protect his legacy. Johnson had confided in him that he would not be seeking a second term in the White House. He was utterly demoralised by his failures in Vietnam and crushed by the hatred he now inspired among so many people, especially the young. ('Hey! Hey! LBJ! How many kids did you kill today?') Webb had shown his loyalty. 'I'm going to walk out the day you do. Let's go out together.' As for Robert McNamara, the one man in all of Washington whose hard-driving aggression had come anywhere near threatening Webb's self-confidence – well, even he was broken now. The 'walking IBM computer' who had spearheaded the escalation in Vietnam had changed his mind about the whole enterprise, but it was far too late. He was tense and gaunt, his former confidence all but destroyed. Occasionally, visitors to McNamara's office would find him with his face turned to the wall to hide his tears.

The auguries weren't good for the Democrats as they headed towards the 1968 election, but whatever the political flavour of the next president, Webb couldn't expect the kind of support he had enjoyed throughout so much of the 1960s. Anyway, he was deeply tired. His plan, then, was to find and groom a non-partisan successor who might be acceptable to a future administration. He could then quietly step down and allow that successor to take credit for the final triumphant landing on the moon. No new president would dare to replace a NASA administrator on watch during that mission, or in the immediate lead-up to it. A new NASA chief would be 'safe' throughout 1969 and beyond, just so long as Apollo delivered the goods 'before this decade is out'. Webb settled on his new deputy Tom Paine as his successor, knowing that Johnson would approve. Once again, it was important that this idea seemed to originate from the White House and not from Webb.

Meanwhile, events in space were accelerating. On 14 September, the Russians finally succeeded in swinging an unmanned Zond/Soyuz capsule around the moon and returning it to earth. Two days later, Webb went to the White House to talk to Johnson about this, and to consider NASA's future. It was time now, he felt, to talk in more detail. He planned to keep an eye on the Agency for another few weeks, but it was best that Paine replaced him well ahead of the handover of political power to the new presidential administration, whether Hubert Humphrey's or Richard Nixon's, after the coming November election. Webb also wanted to stress that he wouldn't be a liability in the election if Johnson chose to campaign for the Democrats. 'If the same guys that had tried to burn him with Bobby Baker and Fred Black and all the rest of it got going again, I wanted to be on the outside where I could take them on in their own terms, and not as an administrator on the defensive.'

Johnson agreed this made sense.

'Alright, I'll make the preparations', Webb said.

'Well, if we're going to do this, we'd better make the announcement.'

'You mean today?'

'Yes. Let's get the press in here.'

'Look, I haven't called my wife –'

'This is the time to do it.'

Johnson seemed in a terrible rush to secure Webb's resignation as soon as possible. Perhaps he was tired of the scandals surrounding NASA in recent months. More likely he just wanted the world to see that he was still the one who controlled the comings and goings of a presidential appointee. Johnson suggested that Webb's 62nd birthday on 7 October would be the ideal time for him to step down. And so the announcement was made. Just as Webb had feared, his wife Patsy heard the news second-hand from a neighbour and tried to reach him at his NASA office. She spoke to Paine, asking him what was going on. Paine was just as shocked as she was. Johnson had been distracted, even insensitive, yet Webb bore him no grudge. Three weeks later, when Patsy Webb arranged a birthday party for her husband, Lyndon and Lady Bird Johnson turned up to wish him well, and Johnson was clearly delighted that Patsy had invited him.

There was another reason to celebrate. Even as the party got under way, Apollo 7 was safely in orbit with a crew aboard: Don Eisele, Walter Cunningham and mission commander Wally Schirra. All three were irritable with head-colds, and their bad temper during the mission might explain why none of them were assigned to fly in space again, but at least the redesigned spacecraft was working properly.

As NASA's new administrator, Tom Paine quickly showed his courage. On 12 November he held a press conference to announce that the next mission, Apollo 8, would be going all the way to the moon.

Discovering the earth

At 7.51 am EST on 21 December 1968, Apollo 8 took off, and this time, as its engineers had so confidently predicted, the Saturn V worked flawlessly. At 9.41 am, mission controllers told Borman and his crew: 'You are "go" for Trans-Lunar Injection.' The Saturn V upper stage fired its engine again, and Apollo 8 blasted out of orbit, heading for the moon. For the first time in history, humans were leaving the protective realm of earth and travelling into deeper space. They were also moving faster than anyone else had ever done, climbing away from the earth at 25,000 mph. Apollo 8 was the 26th crewed spaceship to orbit the earth, and the first to travel onwards to another world.

Next day, 50,000 miles from home, the astronauts sent back TV pictures of the earth: not just the curved horizon but the entire ball of our planet, surrounded by murky blackness. The mono TV camera aboard the ship couldn't make out much more than a white blob, but it was still sensational. Jim Lovell thought out loud for his TV audience: 'I keep imagining I'm a lonely traveler from some other planet, and I wonder what I'd think about the earth from this distance. Would I think it was inhabited?' A few minutes after this TV transmission, Apollo 8 quietly slipped out of the earth's region of gravitational influence and into the moon's.

Borman was a tough commander, more concerned with getting the mission right than with its broader poetic significance. At one point he told his crewmates to stop daydreaming and focus on their instrument panels. 'I don't want to see you guys looking out the window.' When Lovell accidentally inflated the life vest in his survival harness, Borman glared at him. But when Borman wasn't looking, Lovell and Anders took sneaky glimpses out of the window at the dwindling earth. They were amazed and disturbed that they could block their entire home planet from view just by holding up a thumb.

After a flight of 66 hours, the ship was a quarter of a million miles away, and skimming around the moon. Now came a really smart example of mission management. Apollo was supposed to head round the far side of the moon, turn its tail in the direction of flight and fire its large rocket motor to brake the ship into orbit. If the rocket didn't work, then the ship would simply swing around the moon and head back towards the earth without needing any rocket burn. This essential safety feature was called the 'Free Return Trajectory'. In the event, the braking burn went smoothly, although it was an eerie moment when Apollo 8 slipped out of the earth's view and 'behind' the moon, losing radio contact with mission control for an agonising 45 minutes.

On Christmas Eve, most of the world's TV stations carried what was to become a famous live broadcast from Apollo 8, when Frank Borman read the first paragraphs from the Genesis chapter of the King James Bible, and ended the transmission with a poignant message. 'Merry Christmas, and God bless all of you – all of you on the good earth.' Apollo 8's arrival at the moon was a welcome delight after an appalling year on earth.

By now the crew was starting to worry about the second engine burn that would fire them out of lunar orbit and propel them towards home. Would the engine work a second time? Lovell tried to lighten the mood by saying: 'Hey, who'd have believed we'd be spending Christmas around the moon?'

Anders replied drily: 'Just so long as we're not still here come New Year's Day.'

Well, they made it back, having spent twenty hours in lunar orbit, and the world celebrated their achievement. The irony is that Apollo 8 brought back something far more precious than NASA's scientists could ever have imagined. Colour photographs of the beautiful earth rising above the harsh lunar horizon were printed in every major magazine after the mission, and humanity began to appreciate, perhaps for the first time, just what a tiny, fragile, lonely planet this is, in the vast blackness of space. Bill Anders observed later: 'We flew all the way to explore the moon, and the most important thing we discovered was the earth.'

Low profile

In his retirement, Lyndon Johnson preferred to stay within the confines of his beloved Texas ranch – although on 16 July 1969, he did attend the launch of Apollo 11, on its way to the moon to make the first touchdown. It was a sweltering day. Johnson and other not-so important guests sweated in the bleachers. 'It was worse than I thought it would be. I hated it. I hated people taking pictures of me when I felt so miserable.' He was so distracted, and so far removed from the senior VIP enclosures, he didn't even notice that President Nixon had declined to attend, dispatching Vice President Spiro Agnew in his place. (Advisors had warned Nixon that he should stay clear in case of a high-profile launch failure.) Al Rockwell was there in the thick of things, an instant aerospace pioneer now that his acquisition of North American Aviation had lent him, a former axle-maker from Pittsburgh, this technological triumph by proxy. Jim Webb, however, watched from the sidelines of the bleachers. He was being careful not to attract attention away from Tom Paine.

The Apollo 11 mission took off amidst a general feeling of euphoria, and the rest, as they say, is history.

It must have been welcome when, on 5 August, after Neil Armstrong and Buzz Aldrin had performed their long-awaited dust-scuffling antics on the lunar surface and returned safely home, a conciliatory letter from Apollo's old adversary Jerome Wiesner appeared in the *New York Times*:

> *As we celebrate the successful accomplishment of our great lunar adventure and feel pride in our technical achievement, we should give proper credit to James Webb, whose organizational skill, vision and drive played a major part in its success ... It required*

great leadership to motivate the NASA teams and to sustain Congressional and public support for the program … we should honor the man who directed its accomplishment.

Part Seven
After Webb

It's common to suggest that the burgeoning crisis in Vietnam somehow 'caused' the end of the space age and the demise of Webb-style technocratic Big Thinking. Certainly it contributed to a popular sense of distrust in large-scale technological enterprises, but the counterculture campaigners and peace-promoting liberals who so much hated the Vietnam war and feared the technocracy represented by Webb's NASA were not the people who assumed the greater measure of power in America after 1968. The true pull back of Big Thinking wasn't just a liberal backlash against technology. It was a grassroots Republican backlash against what they perceived as the over-reaching of federal government into the nation's affairs.

James Webb's successor Tom Paine found the climate under President Richard Nixon much less amenable than in Webb's day, and he wasn't emotionally ready to accept that NASA had to change. He thought that the lunar landing missions had set a precedent for further large-scale space projects, but he was wrong. Apollo had been the exception, not the rule. It was Paine who was actually forced to call a halt to the moon programme, cancelling missions 18, 19 and 20, and trying to save money and appease the new president. However, he did successfully salvage a working Saturn V for use in the last vestige of the Apollo Applications programme, the Skylab orbital workshop of 1973, America's first space station.

Paine acknowledged that American taxpayers would no longer tolerate writing off expensive Saturn rockets after each flight, but he believed that NASA had a solution to this wastage. The proposed 'space shuttle', championed by George Mueller, was designed to be as reusable and easy to operate as an aeroplane. Paine hoped that Nixon's administration would allow the savings from cancelled

Apollo flights to be allocated for development of the shuttle. Unfortunately, Paine was no match for Nixon's advisors, and he resigned after just two years. 'I finally left because I didn't think I could deliver the kind of relationship with the President which the head of NASA really ought to deliver.'

NASA's next administrator, Jim Fletcher, campaigned for $10 billion to build the shuttle. Under political pressure he revised his estimate to $8 billion. This was still too expensive for the White House. Nixon's advisors asked for another and yet lower budget to be submitted for their approval. Caspar Weinberger, a senior figure in the Budget department at that time, intervened to persuade Nixon that NASA's requests couldn't be shelved indefinitely. Some kind of commitment had to be made. On 12 August 1971, he sent Nixon a memo, a single short document that probably saved the space shuttle from total cancellation. He argued that there was a danger of giving the impression that 'our best years are behind us, and that we are turning inward'. He further stated: 'NASA's proposals have some merit … The real reason for sharp reductions in NASA's budget is that we cut it because it is cuttable, and not necessarily because it is doing a bad job.'

Nixon was in a difficult position. He couldn't count on public support for a big NASA programme, but nor did he want to kill the space effort altogether, because that would have been damaging to America's international prestige (which mattered a great deal to him, since he prided himself on his agenda for foreign diplomacy). So he scrawled a brief note in the margins of Weinberger's memo: 'I agree with Caspar.' That must be one of the most carelessly off-hand authorisations ever written for a major space programme.

However, by January 1972, when Nixon formally announced the shuttle programme, the permitted budget had shrunk to $5.5 billion, a 30 per cent shortfall in NASA's 'lowest option' funding, even before the first pieces of hardware were built. NASA was forced into devising a fudged system with a throwaway fuel tank and twin solid rocket boosters, instead of their ideal design, an elegant 'piggyback' vehicle with a winged booster that could land and be reused just like the main spacecraft itself. A system that could have been very cheap to fly, if only the initial investment had allowed for a proper design, was supplanted by something that was cheap to build and costly to fly. It was a dangerously imperfect piece of technology.

Everyone at NASA had imagined that the shuttle would be just one element of that old dream, an orbiting infrastructure. What was the point of a shuttle if it had nowhere to go? Many detailed space station designs had been formulated over the years, but under the Nixon regime, NASA could only gamble on proving the shuttle concept first, and then winning funds for a space station at some future date.

Confronted by harsh financial realities, Jim Fletcher introduced what seemed at the time a bold and practical initiative, a plan to make the shuttle available as a general-purpose 'space truck' for private industry as well as for NASA scientists. NASA would attempt to offset a proportion of the costs by charging fees for launching other people's payloads. Fletcher reinforced the fragmented political support for a continued manned space programme by promising cheaper access to space for private communications satellites, university experiments and military payloads. This was certainly an appealing prospect for Nixon, and for subsequent presidents. It seemed a way of making NASA a genuine national resource instead of just an expensive club for space cadets. Critics argued, with some justification, that there was no reason to carry astronauts and life-support equipment in the same vehicle as an unmanned satellite, which could always be launched more cheaply in a small expendable rocket. But Fletcher suggested that the shuttle would represent a perfect compromise between manned and unmanned operations, a 'do-it-all' spaceship to suit all purposes. In the event, he was wrong.

Fletcher relied on McNamara-style cost-benefit analyses to support his arguments. He hoped entirely to eliminate NASA's reliance on expendable rockets to put payloads into space. But the shuttle hadn't even been built yet, and nobody could properly claim any previous experience operating such a machine, because nothing like it had ever flown before. Inevitably, all the analysts' figures about its anticipated cheap operating costs and reliability were speculative. NASA approached private companies, universities and research institutes, trying to interest them in buying future payload space aboard the shuttle. At one point, a fee of $220 per kilo ($100 per pound) was predicted, with a shuttle launch rate of 50 or 60 flights per year, divided between a fleet of five orbiters. These hopes would prove to be unrealistic. NASA would achieve less than one flight a month, and payload prices would eventually work out 50 times more expensive than Fletcher's over-optimistic analysts had forecast. (The analysts, by the way, were brought in from the private management consultancy industry. Perhaps they felt they had nothing to lose by telling NASA what it wanted to hear?)

Throughout the 1970s there were several changes of administration in the White House. The Watergate scandal ended Nixon's occupancy in 1974; Gerald Ford's caretaker presidency gave way to Democrat Jimmy Carter, who was subsequently crippled by the Iranian hostage crisis. All these political dramas overshadowed NASA's fortunes. But at last, on 12 April 1981, morale within NASA was revived when the first operational space shuttle, Columbia, blasted into orbit, piloted by veteran astronaut John Young and newcomer Bob Crippen. American astronauts were in orbit again after a furlough of six years. The shuttle seemed to work well. Public approval for space flight reached a new high after a decade of indifference. Best of all for NASA, the latest man in the White House, Ronald Reagan, began to show a keen interest in their new ship. Within his first term in office, he would give his blessing to the next big prize for the space cadets: a giant orbiting station. But first, NASA's chiefs had some hard selling to do.

Island in the sky

Ronald Reagan became President in 1981, sweeping aside Jimmy Carter's beleaguered administration by a comfortable margin at the polls. Reagan's approach to space became the most adventurous since that of President Kennedy in the 1960s. As it turned out, he was thinking very big indeed. On 23 March 1983, he made a startling public announcement: 'Until now, we have based our nuclear deterrence on the threat of retaliation. What if we could live secure in the knowledge that we could intercept enemy missiles before they reached our soil?'

Reagan was promoting the Strategic Defense Initiative (SDI). A network of killer satellites, laser platforms and other exotic weapons would girdle the globe, protecting the Free World against what Reagan once called 'The Soviet Evil Empire'. Edward Teller, inventor of the hydrogen bomb (and the real-life inspiration for the movie character Dr Strangelove) had sold Reagan a pipe dream. Most scientists were doubtful that SDI, 'Star Wars', could be made to work, but military spending under Reagan escalated to levels not seen since the height of the Vietnam War, and Teller's supporters had little difficulty initiating big new projects. A number of high-technology companies made ambitious claims for SDI in order to justify huge research grants.

It might be argued that SDI 'worked', even though it was impossible to prove the effectiveness of its hardware. The Soviet Union couldn't hope to match the research spending, and this had a significant impact on the easing of the Cold War by the end of the 1980s. Reagan may not have been overly concerned with the technical details, but he made effective use of SDI's apparent threat in his subsequent peace negotiations with Russia.

The Pentagon's SDI plans were real enough, however. Vandenberg Air Force Base in California was adapted to handle shuttle launches and landings for defence-related missions – although the military fell somewhat short of paying for a shuttle of their own, as it had once promised. Instead, NASA was persuaded to allow military access to flights at short notice in return for some degree of political support. Civilian space staffers were unhappy with this arrangement. A significant proportion of shuttle flights was now committed to carrying secret military payloads. In the shuttle's first decade of operation, for instance, the vehicle flew successfully on 42 occasions (a rate ten times slower than Jim Fletcher's analysts had hoped for). Details about the payloads and activities for nine of these flights were classified.

In June 1981, the new Reagan administration put James Beggs in charge of NASA. During the 1970s, he had served under President Nixon as Secretary of Transportation. As NASA chief, he set his sights on a space station. He argued that the Russians were still years ahead in terms of earth-orbital experience. He tried to persuade the Air Force, and the Department of Defense in general, to take an interest in the station. He also suggested to the CIA that such a large orbiting platform might be suitable for spying. He was trying to capitalise on Reagan's current enthusiasm for the strategic benefits of space flight. But the CIA's satellite experts were adamant that astronauts walking about inside the station would blur the surveillance cameras. Unmanned systems were cheaper as well as more efficient, they advised.

Meanwhile, Caspar Weinberger, who had done so much to save NASA's shuttle programme from Nixon's cost-cutting in 1971, was unwilling to be so helpful this time. He flatly opposed the space station. As Reagan's Secretary of Defense, he was anxious to preserve as many shuttle flights as he could for defence-related activities. He believed that NASA would swallow up launch capacity and deny his department valuable access to space.

Beggs broke the log-jam by going behind the backs of the station's opponents. He enlisted support from a third party, the Trade and Commerce Council – representatives of all those people within the country apart from NASA who would most stand to benefit from a space station programme: the half-dozen or so giant and politically powerful aerospace corporations; the hundreds of smaller sub-contractors; the tens of thousands of factory workers all across the country; the congressmen and women who represented them and counted on their votes – and on the election campaign funds so kindly donated by the big corporations to politicians who were in sympathy with them ... In the end, even the station's most bitter opponents at the Pentagon would prove no match against the mighty dollar.

On 1 December 1983, Beggs and his allies brought the Commerce people into the White House for a discreet meeting with Reagan. With commercial interests now persuaded over to NASA's side, Reagan used this session as a cue for approval. He announced the space station project in his State of the Union Address on 25 January 1984: as high-profile an occasion as Beggs could have hoped for. 'Tonight, I am directing NASA to develop a permanently manned space station, and to do it within a decade ... We want our friends to help us meet this challenge and share in the benefits. NASA will invite other countries to participate, so we can strengthen peace, build prosperity and expand freedom for all who share our goals.'

Beggs had achieved a significant milestone: presidential approval for a big new manned space project. In the process, he had upset some very powerful people. Reagan's principal science advisor, George Keyworth, had come to the White House with a background in atomic weapons research. He wasn't even remotely supportive towards NASA and its spaceships. Worst of all, Keyworth had been entirely bypassed by Beggs, and was furious that space station plans had been shown to Reagan without his knowledge or approval.

In 1962 a British observer, we may recall, watched Jim Webb clash with Kennedy's science advisor Jerome Wiesner. 'Who will win?' he asked. And Kennedy replied: 'Webb will win. He's got all the money, and Jerry only has me.' Twenty years later, NASA's chief would find the shoe was on the other foot. Keyworth now had all the money and Beggs had only the President. Stubbornly unwilling to acknowledge the danger presented by some of his Washington opponents, he would soon have his entire career and reputation at NASA thrown into doubt, probably as the result of political malevolence directed towards him.

The station itself, and the back-door approval process that Beggs had engineered, were by no means the only matters to upset the apple cart. President Reagan's speech had contained yet another shock for NASA's enemies. Shortly before his State of the Union speech, Reagan had alerted European, Japanese and Canadian leaders of his intentions to invite their participation, sending them private memos. His public announcement on 25 January came as a surprise to many people in America itself. How could Reagan have made such a visible commitment to foreign cooperation without anyone knowing in advance? Beggs's team (and especially NASA's hard-working experts on international affairs, Kenneth Pedersen and Peggy Finarelli) had managed to incorporate an international element into Reagan's speech, with very little external consultation. In other words, they hadn't given their critics a chance to disagree. The prize of the gamble, as they well understood, would be a layer of protection for the station against Congressional cost-cutters. Many politicians would draw back from attacking the project if America's international prestige were seen to be at stake. In the weeks following Reagan's speech, the National Security Council sent Beggs on a tour of Europe, Canada and Japan, to seek space station collaboration. With Reagan's approval, he was given the use of a presidential jet, reinforcing his image as an official envoy.

And so, the space shuttle and the International Space Station were born out of a kind of desperate politics. One cannot say with any certainty how Jim Webb might have handled these projects, but it seems reasonable to speculate that he would have argued, bullied and cajoled all around, from the uppermost levels of the White House to the smallest contractor, for answers to the following questions:

What does America, as represented by the President and by Congress, wish to achieve in space in the foreseeable future?

What actual missions are required to be accomplished?

What precise design of hardware is best suited to achieve those missions, given the budgets that are most likely to be available?

What milestones can we expect to achieve, and in what time-frame, so that budget appropriations equate, at least approximately, with NASA's promises?

NASA's problems with the space station stemmed from the

subversion of the order in which those questions were addressed. First, the station's planners came up with undefined hardware. How could it be defined, when its mission had not yet been specified? Then, missions were created to try to justify the hardware. Finally, political consensus was sought for building the stuff. Since the missions were still vague, they remained open to political argument. Therefore, the hardware designs kept changing in order to adapt to ever-changing missions, and costs consequently rose to such an extent that yet more hardware variations were required in order to stay within achievable budgets. Mission requirements now had to shift yet again, in accordance with the reduced capabilities of the hardware, and so on and so forth, in a terrible spiral of confusion and escalating costs. Sound mission planning got lost in the maze.

The space station contractor companies competed fiercely for work, forging separate and sometimes cluttered alliances with rival NASA field centres. What emerged was a concept called the 'Work Package' in which each of the centres would be assigned portions of the station to develop, while managing their own teams of contractors. But defining which portion belonged to what centre led to many difficulties. The Marshall Center, with its experience of building large structures such as the Saturn V and the space shuttle's fuel tanks and boosters, wanted a gradual assembly of the station, starting with the truss beams and adding modules and manned capability over time. This was known as the 'outside-in' approach, and it reflected Marshall's relative lack of experience in supporting astronauts once they were in space. The (by now renamed) Johnson Space Center in Houston, with its greater emphasis on astronaut activities, wanted the opposite construction regime for the station: manned capability as soon as possible, starting with the modules and adding trusses later. Naturally, this was known as the 'inside-out' mode.

Jim Webb had insisted that all the field centres should be accountable to NASA's space programmes. Now those programmes were being distorted to match various field centres' desires and priorities: a complete inversion of the disciplines that Webb had imposed. NASA was fragmenting, once again, into a disordered rabble of scattered laboratories. Both the shuttle and the space station began construction before their missions were specified – yet another inversion of good sense. Apollo was mapped out on paper, the LOR debate firmly settled, and von Braun's field centre at Marshall persuaded to sacrifice EOR long before the first metal was cut to build the Apollo spacecraft.

Clashing cultures

Marshall in Huntsville and the Johnson Center in Houston couldn't even agree what the station was for. Johnson, with its experience of controlling moon flights and shuttle missions, wanted to use it as a base for continued exploration of space, a classic 'staging post' alongside which future lunar and interplanetary ships could be built in orbit. Even when this notion became unrealistic because of NASA's general budget restraints, the Johnson Center wanted at least to retain this capability as a long-term option. But Marshall, with Skylab on their list of achievements, favoured a more conservative scientific and experiments-based platform.

The field centres were becoming increasingly ungovernable. After so many years of providing jobs and contracts in their respective states, they had secured considerable bargaining power with their local Congressional representatives. This was corrosive to NASA as a whole. For instance, Washington headquarters might instruct Marshall to make a programme reduction, only to be confronted by a delegation of Alabaman politicians threatening to generate a fuss in Congress on Marshall's behalf; while the Johnson Center always benefited from the traditional strong political and financial influence of Texan lawmakers in Congress. These field centre tactics seemed undignified to many outsiders, who still tended to see NASA as the more cohesive entity that Webb had forged during the Apollo years. The space programme as a whole gained a new and unwelcome reputation as fractious and poorly managed.

What was needed was firm management from headquarters. Unfortunately, by the winter of 1985 the place was in chaos. Beggs's deputy was Bill Graham, a White House political appointee sponsored by Reagan's chief science advisor, George Keyworth. A legal problem stemming from Beggs's previous commercial career forced him to take a 'leave of absence' from NASA's top job. (John Logsdon says that Beggs was 'sandbagged by the White House'.) Graham became acting administrator, but Beggs continued to run an office from inside headquarters. For a while, a gruesome situation developed in which Beggs and Graham appeared to be running rival fiefdoms within the corridors of NASA. Beggs felt betrayed by the White House, and wondered if his legal problems had been created by political schemers to edge him out of NASA in Graham's favour. Eventually, every charge against him was dropped for lack of evidence. This particular feud gives some flavour of the many tensions preoccupying senior NASA managers at the time.

Graham wasn't a rocket expert, but his management instincts were sound enough. Towards the end of November 1985, he noticed some odd technical discrepancies in documents analysing the reliability of shuttle rocket motors. Genuinely disturbed, he made a note requesting better data. But information flow within the Agency was log-jammed at all levels. Middle-ranking managers found their allegiances shifting this way and that between Graham and Beggs: a circumstance that generated dangerous confusions. Morale at all levels was dampened, not just by the leadership crisis, but also by the increasing pressures of the shuttle schedule.

Challenger

On 28 January 1986, at 11.38 am, space shuttle Challenger lifted off into a freezing winter sky. Seventy-three seconds into the flight, it blew apart. All seven crew members were killed: Commander Dick Scobee and his co-pilot Mike Smith; mission specialists Judy Resnick, Ellison Onizuka, Ron McNair and Greg Jarvis; and a high-school teacher launched as part of a schools' education programme, Christa McAuliffe.

Today we are in a position to compare and contrast the 1967 Apollo accident with the recent losses of two space shuttles. The people of Apollo were caught off-guard by something never encountered before, a dramatic fire in a spacecraft sitting quietly on the ground. The shuttle disasters were flagged repeatedly by earlier failures: partially burned rubber seals in boosters, and incidents of chipping and cracking in the heat shields. Taxpayers and politicians are willing to forgive such disasters, but only if NASA can be said to have acted in good faith by trying to prevent them.

After the Challenger accident, the renowned physicist Richard Feynman was asked to join the board of inquiry (the 'Rogers Commission'). Frustrated by the dull and pointless verbiage of due process, he staged an unauthorised little drama for the TV cameras that forced the pace. He dipped a piece of rubber into a glass of iced water to show how it hardened when cold. 'Do you suppose this might have some relevance to our problem?', he asked, knowing very well that it did. He'd created a vivid demonstration of the notorious 'O' ring problem that had destroyed the shuttle during a freezing winter launch.

Feynman wanted to probe further. 'If NASA was slipshod about the leaking rubber seals on the solid rockets, what would we find if we looked at the liquid-fuelled engines and all the other parts that make up a shuttle?' He was told the inquiry wasn't briefed to look at those, because no problems had been reported; so he made unauthorised trips to several NASA field centres and did a little digging. As a charming and sympathetic physicist, he had little difficulty winning the trust of practical engineers at ground level. 'I had the definite impression that senior managers were allowing errors that the shuttle wasn't designed to cope with, while junior engineers were screaming for help and being ignored.' He'd hit the nail on the head, but he'd done it out of school. Finally, he turned to the complex relationships between the space agency's various departments and their suppliers: 'NASA's propulsion office in Huntsville designs the engines, Rocketdyne in California builds them, Lockheed writes the instructions and NASA's launch center in Florida installs them. It may be a genius system of organization, but it seems a complete muddle to me.' In the last days of the inquiry, he made a plea for realism. 'For a successful technology, reality must take precedence over public relations, because Nature cannot be fooled.'

Other commentators identified a malaise within NASA, wherein a bureaucracy that was once created to get a specific job done became, instead, an organisation whose only purpose was to ensure its own survival. That crisis stemmed from a lack of strong, committed leadership in the setting of goals, both from within NASA and the White House. If the people in an organisation have a goal to believe in, they will work effectively and cooperatively. If that goal is lacking, then people start worrying about simply keeping hold of their jobs. They will compete internally with each other, and the organisation will fail.

Back to the moon?

The loss of a second space shuttle, Columbia, in February 2003 highlighted that NASA could no longer be allowed to drift into chaos. It needed a sense of direction. On 14 January 2004, President George W. Bush made a televised speech during a special visit to NASA headquarters in Washington. 'Today I announce a new plan to explore space and extend a human presence across our solar system.' The first goal, he said, was to get the shuttles flying again and finish the space station. No surprises there. But as his speech continued, radical new ideas emerged. 'Our second goal is to develop and test a new spacecraft, the Crew Exploration Vehicle, by 2008, and to conduct the first manned mission no later than 2014 ... Our third goal is to return to the moon ... Using the Crew Exploration Vehicle, we will undertake extended human missions to the moon as early as 2015, with the goal of living and working there ... With the experience and knowledge gained on the moon, we will then be ready to take the next steps of space exploration: human missions to Mars and to worlds beyond.'

These plans emerged after months of secret consultation between White House officials and space experts during 2003. Bush had personal reasons for being cautious. Back in 1989, his father also promised a mission to Mars. NASA presented an overblown $400 billion budget that appalled everyone, and the idea was scrapped almost immediately. The son didn't want to repeat his father's mistake, and he made only vague references to a Mars mission in his 14 January speech. But when it came to the moon, he was more confident about possible dates. After all, NASA already knows how to get there. It should be affordable, and it's only three days' flight time away.

Bush's timing was influenced by presidential elections coming up in November 2004. The Iraq crisis had defined his presidency to the exclusion of almost everything else. Like Kennedy before him, he wanted to be celebrated for something more up-beat than war. 'We choose to explore space because doing so improves our lives and lifts our national spirit', he said. In a strange echo of Kennedy before him, he hoped that a bold announcement about space could divert attention from some of his policies on the ground.

Drastic changes will have to be made within NASA if Bush's vision is to become real. Construction of the space station will end in 2010, so that funds can be diverted towards the new lunar project. Russia, Europe and Japan will then be responsible for the station. At the same time, the space shuttle, that doomed, dangerous winged monster, responsible for many successes and two appalling tragedies, is to be retired. After 2010, the 'Crew Exploration Vehicle' (CEV), highlighted in Bush's speech, will become America's main space vehicle.

The CEV is a reversal of everything NASA has worked towards over the last 25 years. The dream of an all-purpose shuttle, flying cheaply and regularly like a commercial cargo plane, never came to fruition. Whenever a shuttle puts a satellite or a space station module into orbit, the costs are huge because astronauts always come along for the ride. The launch weight given up to life support and crew cabins has to be subtracted from the cargo, and that limits how much payload a shuttle can carry. It's more efficient, say the CEV's designers, to haul cargo in unmanned rockets, and launch astronauts separately in a small module that carries people and nothing else. There's no need for the shuttle's huge wings and cargo bay doors. All that's left to worry about is the crew cabin at the front. Replace that with a cone-shaped capsule that fits on the end of a conventional rocket, and it can also serve as an escape pod if anything goes wrong during launch. And that's basically the CEV. It's a highly adaptable component in a 'building block' system. It plugs onto booster rockets, fits on top of lunar transfer modules and landers, and it can even dock with the space station.

Is this forward progress, or a weird leap back in time? Robert Zubrin has worked tirelessly over the last decade to promote a human Mars mission, and he's not sure. The topsy-turvy logic of space planners often exasperates him. 'In the 1970s, when President Nixon killed the Apollo program and ended lunar exploration, NASA said we would do all that again one day, after we had developed cheaper transportation to orbit using a winged shuttle. We've spent 25 years trying that, with no positive results. The shuttle costs more to fly than the Saturn boosters we had for the Apollo missions. Now the shuttle is going to be replaced with a CEV capsule just like Apollo.'

Bob Seamans was one of the NASA veterans brought aboard to advise the CEV designers. 'I served on what they called their "Graybeard Committee". All these old hands who knew how we'd reached the moon first time round. Astronaut John Young was there, and he'd flown in Gemini, two Apollo missions and commanded the very first shuttle flight. We didn't fool around. They put us in a room from eight till five. They brought in food because there was no break for lunch ... What they've come up with is amazingly similar to what we had with the Apollo.'

After more than 30 years, it's impossible to recreate Apollo's giant Saturn rockets and gantries. All the factory tooling was scrapped, and even the original drawings have been lost. Instead, the CEV will exploit a smaller rocket currently in use, the Delta IV satellite launcher. It's powerful enough, especially if CEVs and other modules, launched on separate rockets, are docked and combined in orbit. Another option is to adapt existing shuttle hardware, and especially the external hydrogen tanks and solid boosters. The winged orbiter could be replaced with a simple payload shroud for CEV hardware. But it's probable that NASA will abandon shuttle hardware completely.

The Columbia Accident Investigation Board (CAIB) decided that it's too risky to carry astronauts in the same part of a spacecraft that contains the rocket engines, because of the risk of explosions. It's wiser to carry the crew in a separate pod which can be blasted away from a malfunctioning booster in the event of problems during launch. The Board noted that the Apollo capsules were extraordinarily safe once the fire hazards had been ameliorated. When Apollo 13 blew up on the way to the moon in April 1970, the rear service module with the rocket engine was torn wide open, but the command module itself was unharmed and returned safely to earth. The shuttle has no such distinction between its crew cabin and the rest of the system. This has cost fourteen astronauts their lives, and will always be a problem for any space plane that doesn't have a separate escape module. A return to tried and tested technologies may be the best way forward for NASA's human space flight programme.

The new lunar missions haven't been defined, but in principle, the CEV will dock in earth orbit with an unmanned transfer vehicle, launched separately: the apotheosis of Wiesner's beloved Earth Orbit Rendezvous. The combined ship will then fly to the moon. The CEV will detach and drop down to the surface, using a small landing stage. At the end of a mission it will be boosted back into lunar orbit by a yet smaller ascent stage. Apollo's lunar module lives again, albeit on a slightly larger scale. Similar hardware will deposit moonbase components ahead of the astronauts' arrival. Inflatable yet strong living quarters, already developed for the space station (but never launched because of political interference from the builders of existing metal modules) are being adapted for the moon. It won't be difficult to set up a base. NASA has been studying how to do this for many years.

But is the moon the right place to go? Do we need a human presence there? According to Bush and his supporters, we must gain experience of living on the moon before we even consider doing the same on Mars. And anyway, the moon is a worthwhile target in its own right. One priority may be to build a super-sensitive astronomical telescope inside a crater, shielded by the moon's bulk from the electromagnetic 'noise' of the sun and earth. The other big prize will be mineral wealth. If there's anything under the lunar dust worth a buck or two, NASA wants to find out.

Bush's speech wasn't exactly poetic, but he was right when he said: 'In the past 30 years, no human being has set foot on another world, or ventured farther upward into space than 386 miles, roughly the distance from Washington, DC to Boston, Massachusetts. America has not developed a new vehicle to advance human exploration in space in nearly a quarter-century. It is time for America to take the next steps.' The hardware for that next step will be designed soon enough. The bigger challenge will be to maintain political support over the coming decade. It is already wavering. Someone with the tenacity of James E. Webb might be able to draw down the moon once again. But perhaps they don't make people like him any more.

Epilogue

On 19 July 1979, a decade after the Apollo 11 landing, Margaret Chase Smith wrote an impassioned ceremonial letter to be lodged with the Library of Congress. It stands as a fine testimony to James E. Webb's years at NASA. It is also a rebuke, on permanent record, for a nation whose gigantic achievements in science and exploration are often overlooked by its apathetic and easily bored broader culture:

> *I am disappointed and concerned with three matters. The first is the public apathy on the space program. Once we had a man on the Moon, the American public lost interest ... My second disappointment was the inevitable consequence of public apathy. It was merely the next step. It was the decline and fall of the Senate Space Committee. When the public lost interest, its elected representatives in Congress found the space program less productive on providing votes for reelection. There was so little interest in the Committee that in 1973 the Senate Majority had to go outside of the Committee to get someone to agree to chair the Committee as none of the sitting members would agree to becoming Chairman. ... My third disappointment has been the lack of recognition for the man who put a man on the Moon – for James E. Webb, the Space Administrator during its most crucial years ... He had to take the heat and fire of partisan political attacks from headline-hungry politicians ... I saw this first hand in my work on the Senate Space Committee! But as compared to the hero astronauts, what recognition or material gains did Jim Webb, and the thousands behind the scenes he typified, receive? Minimal, if any. They declined to commercially exploit their official positions. And today, they are forgotten men and women of a forgotten program ... What we*

> *need today on the energy crisis is another Jim Webb, who not only could build and inspire a crash program team, especially on solar energy, but who would have the credibility and rapport with the Congress on getting cooperation on necessary legislation and getting on with the crucial job needed to be done.*

'The energy crisis' Smith was referring to was the oil price escalation of the early 1970s, which sent many Western economies into a tail-spin. Three decades later that issue remains one of our most pressing concerns. Who, then, will manage the necessary solutions? Who will take charge? Leadership can be a terrifying threat to the freedoms of people beneath the chain of command, or it can be the knife which cuts through confusion to the common good. Webb's case is fascinating, for his career exemplified the drawbacks of large-scale governmental manipulation, even as it showed the wonders that could be achieved when thousands of people worked in harmony to solve a particular challenge.

And on that note, it must be obvious by now that the challenges of Apollo, though complex, were triflingly simple in comparison to the energy problem, global warming, world poverty, ideological wars and all the other ills that afflict us. Webb's space age management was best suited to the space cadets after all, not the rest of us. NASA's moon was a clearly defined target, and if we choose to go back there any time soon, or venture onwards to Mars, then we would indeed be wise to remember how Webb did things and do likewise ourselves. But our social burdens on the ground cannot be similarly quantified in pounds per square inch of payload, and the gulfs that separate different cultures in this new century cannot be measured in nautical miles, nor bridged with any certainty 'before this decade is out'. And what if we do identify some great task that can best be accomplished with machines? It's no longer obvious that governments should have any central role in creating the technology. Half a century ago, wise advisors might have been able to identify potentially useful areas for large-scale national research, such as aviation, computing, rocketry and nuclear energy. Today, it's all anyone can do to just keep in touch with the bewildering pace of developments in medicine, genetics, electronic consumer goods, personal computing and global communications. The politicians are hard-pressed merely to cope with these myriad devices, let alone urge their invention.

Even the bombs and missiles that were once our darkest pride have lost their edge in an era of 'asymmetrical warfare'. A few fanatics with dime-store craft knives can change the world in a day, while the ICBMs and their costly megatons stay sealed in their silos, impotent in a world that barely even worries about them any more. As for space exploration, that ambiguous child of an antique and near-forgotten Cold War: we have reached a time when NASA is no longer the proud vanguard of a new technology. It may instead be the faltering custodian of an old one. The significant legacy of Webb's NASA is not technological but moral. He and his federally employed colleagues were unafraid to accept responsibility for their appointed tasks, whether in luminous success or scalding failure. This lesson, above all, needs to be retained from the lunar adventure. As we send Apollo's machines to the museum, we should keep alive the memory of its people.

‘What do you hope to accomplish?’

After leaving NASA, Webb made sure to keep well away from aerospace affairs so that none of his successors at NASA would fall under his shadow. He practised law in Washington, joined the board of the McGraw-Hill publishing company and spent much time writing and lecturing. The mood in the country had changed, and few people in the 1970s still wanted to listen to him on the subject of managing large-scale technologies. His main professional appointment was as a senior regent of the Smithsonian Institution, a post he accepted at the direct invitation of Congress. And here he did have some notable success, overseeing a tripling of the Smithsonian's budget and witnessing the completion of one of its greatest new facilities. Just off the Mall in Washington DC, the $42 million Smithsonian National Air and Space Museum was constructed to house Chuck Yeager's sound-breaking craft, the X-1, Charles Lindbergh's *Spirit of St Louis*, and (among many other aerospace treasures) the Apollo 11 capsule. The museum opened its doors on the morning of 1 July 1976. A prettily-coloured inaugural ribbon was stretched across the main entrance. President Gerald Ford stepped up in front of it, but he didn't make the crucial cut. A radio pulse from a tiny transmitter deep in space triggered an electric guillotine, and the ribbon was sliced by the very remotest of remote control, using a signal from Viking 1, on its way to a touchdown on Mars. The mission was a headline-grabbing success for the still relatively little-known Jet Propulsion Laboratory in California, whose future Webb had fought to save.

In 1975, aged 69, Webb learned that he had Parkinson's disease, a degenerative nerve disorder that substantially incapacitates its victims. Characteristically determined, he refused to surrender to the disease, although after 1984 he began to work more and more from his home in north-west Washington, with his devoted wife Patsy taking on many of the responsibilities for his care. He accepted numerous awards, honorary degrees and encomiums, as the political establishment and the world of business and academia came at last to see how great his contributions to national life had been. He lived long enough to witness the Challenger disaster of January 1986, and the crippled version of NASA it represented. 'There was an organization that was regarded as being perfect, that suddenly doesn't do the simplest thing', he commented sadly.

Bob Seamans was a regular visitor in those later years. 'We would sit and chat, and pretty soon it was like the old days. Over time [his health] got progressively worse. Sometimes he would be wheeled into the room, or drive himself in a little electric cart ... Invariably, though, he had a report or an article he wanted to show me. "I'd like to know your thoughts on it," he'd say. "Drop me a note and tell me what you think." It was as though I was still working for him!'

When Webb was hospitalised in 1991, Seamans immediately went to see him. His last encounter with his old chief was a strange replay of their parting exchange at NASA more than twenty years earlier. It was extremely hard for Webb to talk, and Seamans had to lean towards him and listen closely.

'Bob, I'm not going to live much longer, but I'd like to know what you hope to accomplish before you kick the bucket.'

James Edwin Webb
7 October 1906–27 March 1992

Notes

Part One: Accepting the Challenge

The PSAC was instructed to come up with a technical briefing about space.
Introduction to Outer Space, 26 March 1958. Reprinted in full in Killian, James R. Jr., *Sputnik, Scientists and Eisenhower*, pp. 288–99.

'Repeatedly, I saw Ike angered by the excesses, both in text and in advertising, of the aerospace-electronics press …'
Killian, *Sputnik, Scientists and Eisenhower*, p. 238.

During the war, a hellish underground factory was built at Nordhausen …
Neufeld, Michael J., *The Rocket and the Reich: Peenemünde and the Coming of the Ballistic Missile Era*, pp. 209–13.

He staged a brilliant escape … under the noses of SS squads …
Cadbury, Deborah, *Space Race*, pp. 37–40.

The National Advisory Committee for Aeronautics (NACA) was established in 1915 …
An attractively illustrated overview of NACA can be found in Gorn, Michael, *NASA: The Complete Illustrated History*, pp. 11–69.

The Pentagon fought a rearguard action to protect its 'latitude to pursue those things [in space] that are clearly associated with defense objectives …'
Killian, *Sputnik, Scientists and Eisenhower*, p. 136.

'I know you're not listening to me because you haven't asked me *how much* you're needed in Oklahoma …'
Duane Roller, Curator of the History of Science Collection at the University of Oklahoma, recalled this story, as told to him by Webb. Quoted in Lambright, Henry W., *Powering Apollo: James E. Webb of NASA*, p. 71.

'In the Congressional wheeling and dealing to land juicy contracts for their home states, all members are equal, but some are more equal than others …'
'Changing Vistas – East, West, South', *Newsweek*, 8 October 1962, 62, p. 27.

'… Apparently they talked about whether I should go into the Treasury.'
Webb interview with John Logsdon, 15 December 1967.

'I don't think I'm the right person for this job …'
Dialogue recalled by Webb in many interviews. Quoted here from his interview with T.H. Baker for the Lyndon Baines Johnson Library, 29 April 1969.

'… I wouldn't take a second-hand invitation to it.'
Interview with Logsdon, 1967. While reviewing the typed transcript a few weeks

later, Webb thought that perhaps the phrase 'second-hand invitation' was too strong. However, he did not ask for this verbatim phrase to be changed.

'I would not have taken the job if I could honorably and properly not have taken it.'
Ibid.

Stocky and voluble, Webb at fifty-five was of a different generation …
Murray, Charles and Bly Cox, Catherine, *Apollo: The Race to the Moon*, p. 70.

'He'd play that good ol' Southern boy "I just fell off the turnip truck" routine …'
Author interview with Roger Launius, June 2005.

'That was one of the best lunches we've had together. I learned so much.'
Murray and Cox, *Apollo*, p. 72 (footnote story).

'The town was full of young men who were riding around in government cars …'
Interview with Logsdon, 1967.

'The way the organization was set up, it was difficult for me to exercise my responsibility …'
Seamans, Robert C. Jr., *Aiming at Targets: the Autobiography of Robert C. Seamans*, p. 73.

'The benefit of the three of us was that we liked each other …'
NASM Oral History Project. Interviewers: Dr David DeVorkin, Allan Needell and Dr Joseph Tatarewicz, 12 April 1985.

'… And they were sort of sticking out their foot to trip us up …'
Interview with Logsdon, 1967.

We should stop advertising Mercury as our major objective in space activities.
'Report to the President-Elect of the Ad Hoc Committee on Space', 10 January 1961, pp. 16–17. NASA History Office.

'… In each case, Kennedy said, "I'm going to stick with you."'
Webb stresses this story in almost all interviews concerning his relationship with Kennedy. This quote from NASM Oral History Project. Interviewers: Dr David DeVorkin, Dr Joseph Tatarewicz and Ms Linda Ezell, 22 March 1985.

'… So it's an interesting question: what would have happened if we had made that original date? …'
Author interview with Dr John Logsdon, June 2005.

Ham appeared to be in good physiological condition, but sometime later, when he was shown the spacecraft …
Grimwood, James M., *Project Mercury: A Chronology*, p. 121.

'… Then, at 900,000 feet, you'll get the feeling that you *must* have a banana.'
London *Daily Mail* cartoon, reproduced in Cadbury, *Space Race*, p. 226.

'"No", the President answered wearily. "Give me the news in the morning."'
Murray and Cox, *Apollo*, p. 76.

SOVIETS PUT MAN IN SPACE. SPOKESMAN SAYS US ASLEEP.
Shepard, Alan and Slayton, Deke, *Moonshot*, pp. 105–06. See also: 'We Are Asleep', London's *The Times*, 13 April 1961, p. 12.

Kennedy stopped again a moment and glanced from face to face. Then he said quietly, 'There's nothing more important.'
Sidey, Hugh, *John F. Kennedy, President*, pp. 121–3.

Then someone asked him, 'Mr Seamans, do you think we ought to put our plans for the moon on a crash basis?'

Author interview with Dr Robert Seamans, September 2005. See also: Seamans, *Aiming at Targets*, p. 86.

Is there any other space program … in which we could win?

This memo is frequently quoted in space history books. Young, Hugo, *Journey to Tranquillity: The History of Man's Assault on the Moon* includes a photographic reproduction of the actual memo in the picture section following p. 136.

Kennedy's response disclosed more than anything the sight of a man obsessed with failure.

Ibid., pp. 108–09.

'Johnson sent the President a report so loaded with assumptions that a moon landing was the inescapable conclusion.'

McDougall, Walter A., *The Heavens and the Earth: A Political History of the Space Age*, p. 319.

'… you'd better doggone well be able to do it.'

Interview with Logsdon, 1967.

'… and you and the President have to recognize we can't do this kind of thing without that continuing support.'

NASM Oral History Project. Interviewers: Dr David DeVorkin, Allan Needell and Dr Joseph Tatarewicz, 12 April 1985.

'We didn't want to spend our time arguing with officials in the Pentagon …'

NASM Oral History Project. Interviewers: Dr David DeVorkin, Dr Joseph Tatarewicz and Ms Linda Ezell, 15 March 1985.

'McNamara … was always an enemy of the Air Force and more a friend to the Army.'

Dyson, George (quoting his father Freeman Dyson) in *Project Orion: the Atomic Spaceship, 1957–1965*, p. 257.

He suggested that NASA should aim for a … trip to Mars instead …

Author interview with Seamans, 2005. See also: Seamans, *Project Apollo*, p. 19.

McNamara would eventually advise the White House that *without* a lunar programme, many thousands of aerospace workers would have to be laid off.

McDougall, *The Heavens and the Earth*, p. 319.

'… I think part of McNamara's concern was just a general feeling, "how could anything as big as this be well-run unless I'm running it?" …'

Interview with T.H. Baker, 1969.

The non-military … 'civilian' projects … are, in this sense, part of the battle along the fluid front of the Cold War.

Often referred to in shorthand as the 'Webb-McNamara Report', this document is actually entitled *Recommendations for our National Space Program: Changes, Policies, Goals*. May 1961. Courtesy of the JFK Library, Presidential Office Files.

'My own feeling … is that our two major organizational concepts … are going to have to be re-examined and perhaps some new invention made.'

Letter from Webb to Keith Glennan, 14 April 1961. NASA History Office.

'I put an administrator's discount on it.'

NASA chief counsel Paul Dembling recalled Webb's phrase, interviewed in Trento, Joseph, *Prescription for Disaster*, p. 49.

'If there is no military value ... and no scientific value ... it will mean we would have ... wound up being the laughing stock of the world.'
Fortune, November 1963, p. 125.

'... We were not just constructing the fastest, quickest way to get a few payloads into orbit. We were building permanently.'
Interview with T.H. Baker, 1969.

Part Two: Bargaining for Power

'I can't do that. When I was at RCA, Bob was junior to me.'
Author interview with Dr Robert Seamans, September 2005. See also: Lambright, Henry W., *Powering Apollo: James E. Webb of NASA*, p. 114.

'... "You go and tell the Congressman or whoever it is putting pressure on you that I told you not to do it, and he should call me."'
The Syracuse/NASA Program Project Manager Research Group, interview with Webb. Interviewers: Henry J. Anna, H. George Frederickson, Barry Kelmachter, 15 May 1969.

This cunningly preserved booster ... under a new name, 'Juno'.
Stuhlinger, Ernst and Ordway, Frederick I. Jr., *Wernher von Braun: Crusader for Space*, p. 128.

'You have to remember that he was "Mr Interplanetary Travel." He was known well by a lot of people ...'
Ibid., p. 231.

'... You've got to be imbued with the idea that a system of accounting is not a foolish thing ...'
NASM Oral History Project. Interviewer: Martin Collins, 4 November 1985.

... no one would doubt Goddard's track record in successfully launching ... a wide range of scientific satellites, including the Hubble Space Telescope ...
Smith, Robert W., *The Space Telescope*, p. 192.

'... But you aren't doing anything about it. The snowball's going to run whether you clap your hands or not.'
NASM Oral History Project. Interviewers: Dr David DeVorkin, Dr Joseph Tatarewicz and Ms Linda Ezell, 22 March 1985.

'Mr Webb, why ... do we have to go to Houston, of all places?'
NASM Oral History project, interview with Dr Robert Gilruth. Interviewers: Dr David DeVorkin and Dr John Mauer, 2 March 1987. See also: Trento, Joseph, *Prescription for Disaster*, p. 42 and Murray, Charles and Bly Cox, Catherine, *Apollo: The Race to the Moon*, p. 130, in which Charles Donlan, Gilruth's deputy at the Space Task Group, says that the Houston decision was 'as though you went through a maze knowing all the time what door you were going to come out'.

'Albert, you know Jim Webb's thinking of putting that new space installation in your district?'
Webb interview with T.H. Baker, Lyndon Baines Johnson Library, 29 April 1969.

'We did what we thought was right for the program, and we let the politicians take the credit when and where they wanted to ...'
Ibid.

Would appreciate article on Texas as background Johnson. Cowboys, oil, millionaires, huge ranches, general crassness, bad manners etc.
Cooke, Alistair, *Talk About America*, p. 81.

'Cowboys there are, west of the Johnson country ...'
Ibid.

John Logsdon, one of America's foremost space historians, is quite blunt when he says, 'It was a real estate scam.'
Author interview with Dr John Logsdon, June 2005.

'You know, some of my fellows here tell me that you're favoring Lyndon Johnson ...'
NASM Oral History Project. Interviewer: Martin Collins, 15 October 1985.

'Kennedy turned around to me and said, "What's this all about?"'
NASM Oral History Project. Interviewers: Dr David DeVorkin, Dr Joseph Tatarewicz and Ms Linda Ezell, 22 March 1985. See also: Swenson, Grimwood, Alexander, *This New Ocean*, p. 470.

Kerner complained bitterly that Illinois's senators 'just won't wheel and deal'.
'Changing Vistas – East, West, South', *Newsweek*, 8 October 1962, pp. 27–9.

JATO's descendants quickly came of age, even as Parsons' wayward career disintegrated ...'
For a fascinating account of JPL's early years, see: Pendle, George, *Strange Angel: The Otherworldly Life of Rocket Scientist John Whiteside Parsons*.

'They have tremendous esprit de corps. It's almost offensive. It's like the Marines.'
Van Allen quoted in Heppenheimer, T.A., *Countdown*, p. 293.

'I thought at one time you were one of the main problems, but I've concluded that we couldn't have done the program without you.'
Pickering quoted by Webb, NASM Oral History Project. Interviewers: Dr David DeVorkin, Dr Joseph Tatarewicz and Ms Linda Ezell, 22 March 1985.

'It was one of the greatest speeches I've ever heard, about the poor scientists ...'
Seamans, Robert C. Jr., *Aiming at Targets: The Autobiography of Robert C. Seamans*, p. 103.

'... they'd all apply to Harvard, MIT and CalTech. Then they'd be turned down ... and they'd feel like second-class citizens.'
NASM Oral History Project. Interviewers: Dr David DeVorkin, Dr Joseph Tatarewicz and Ms Linda Ezell, 22 March 1985.

'Those now in college will before long be living in the age of intercontinental ballistic missiles ...'
Quoted in McDougall, Walter A., *The Heavens and the Earth: A Political History of the Space Age*, p. 161.

'I never realized he had not understood the program that I had laid out for him and put $100 million into ...'
NASM Oral History Project. Interviewers: Dr David DeVorkin, Dr Joseph Tatarewicz and Ms Linda Ezell, 22 March 1985.

'The fact is, there was not a warm, loving atmosphere at JPL for CalTec students ...'
Ibid.

'... the materials used must be of proven permanence. I do not want faded and flaking pictures in a government archive.'
Cooke, Hereward Lester and Dean, James D., *Eyewitness to Space*, pp. 11–13.

'Somewhere around October of 1961, only a few months after Kennedy had decided to go to the moon, there was a terrible argument.'
Author interview with Seamans, September 2005. See also: Lambright, *Powering Apollo*, pp. 119–20 for an administrative overview of the tense relationship between Webb and McNamara.

... an introspective, painfully shy engineer with a scholar's tastes.
'Changing Vistas – East, West, South', *Newsweek*, 8 October 1962, pp. 27–9.

With Atwood's blessing, North American made a furiously committed bid for the contract to build the stage. And on 11 September, it won.
Baker, David, *Spaceflight and Rocketry: A Chronology*, p. 125.

Storms and his troopers drew up a fully detailed bid for Apollo, spending nearly five times their authorised budget in the process.
Gray, Mike, *Angle of Attack: Harrison Storms and the Race to the Moon*, p. 97.

Martin was awarded 6.9, while North American and General Dynamics shared second place at 6.6 ...
'Source Evaluation Board Report: Apollo Spacecraft'. NASA RFP-9-150, 24 November 1961, NASA History Office.

'I just wanted to tell you personally ... You've won Apollo.'
Gray, *Angle of Attack*, p. 115.

By the mid-1960s the company was in a surprisingly delicate position, with profit margins of just 2 per cent ...
'How 300,000 Work For 3 "Moon Men"', *Newsweek*, 21 December 1964, pp. 20–3.

Part Three: Maintaining Momentum

Wiesner happily listened to some enthusiastic presentations in which the direct ascent concept was reanimated.
Young, Hugo, *Journey to Tranquillity*, p. 14.

'Webb's got all the money, and Jerry's only got me.'
Murray, Charles and Bly Cox, Catherine, *Apollo: The Race to the Moon*, p. 143.

'I understand it's a question of whether we need four hundred million dollars more to maintain our present schedule, is that correct?'
This and the verbatim quotes that follow are taken from the transcript 'Presidential Meeting in the Cabinet Room of the White House, November 21, 1962. Topic: Supplemental Appropriations for the National Aeronautics and Space Administration.' Presidential Recordings Collection tape #63. Transcript prepared by Dwayne A. Day, Glen Swanson, John M. Logsdon, Stephen Garber and Robert Seamans. Courtesy of the Presidential Recordings Program, Miller Center of Public Affairs, University of Virginia.

According to Seamans, there was no animosity between the President and NASA's chief that day ...
Author interview with Dr Robert Seamans, September 2005.

Next day's *Washington Post* reported there had been a three-hour brawl with 'a lot of wrangling and raised voices'.
'House Group Grills Webb On Holmes', *Washington Post*, 19 June 1963, p. A2.

'I never got along very well with Webb ... If I'd been a bit more mature, I would have understood how a politician thinks ...'

Stuhlinger, Ernst and Ordway, Frederick I. Jr., *Wernher von Braun: Crusader for Space*, p. 232.

'... But it would take an additional 150 pounds of fuel inside the Saturn rocket at the moment of lift-off to account for getting that half-pound of shaver to the moon ...'

'How 300,000 Work For 3 "Moon Men"', *Newsweek*, 21 December 1964, pp. 20–3.

'Well, the boys came by to see me last night.'

Webb interview with T.H. Baker, Lyndon Baines Johnson Library, 29 April 1969.

'Say, the *New York Times* is after me, and the astronauts apparently have been offered free houses in Houston, and they're after the White House on this thing. How does the White House get into this?'

Ibid.

Today, few people remember much about the ERC, for it was ... a failed dream.

For a detailed account of the ERC, see: Butrica, Andrew J., *The Electronics Research Center: NASA's Little Known Venture into Aerospace Electronics*, American Institute of Aeronautics and Astronautics, AIAA 2002-1138.

'Look, if you don't tell me that this is just the gutter politics of Teddy Kennedy, I'm going to tell you you're a liar!'

NASM Oral History project. Interviewer: Mr Martin Collins, 15 October 1985.

... Lovell reported his discussions with senior Russian scientists in Moscow. Apparently Russia had no plans to go to the moon.

Letter from Sir Bernard Lovell to Hugh Dryden, 23 July 1963, NASA History Office. See also: McDougall, Walter A., *The Heavens and the Earth: A Political History of the Space Age*, p. 395.

The United Nations speech was a political gesture ...

Ibid., p. 394.

'Maybe there were other factors ... There were indications that people around him wanted this to be a slight withdrawing of support ...'

Interview with T.H. Baker, 1969.

'... the space program is probably the most centralized government spending program in the United States ...'

Young, *Journey to Tranquillity*, p. 199. See also: McDougall, *The Heavens and the Earth*, p. 393.

'... So I made it a rule, "We will not ask for a supplemental, Mr Congressman. When you decide you want to make a cut, that's what you're going to live with ..."'

NASM Oral History Project, interview with Webb. Interviewers: Dr David DeVorkin, Allan Needell and Dr Joseph Tatarewicz, 12 April 1985.

'Well, is there anything personal between you?' Kennedy asked ... 'Don't let it get personal.'

Interview with T.H. Baker, 1969.

'I've just got to get ... a tax bill through, and Harry Byrd won't support it ...'

Ibid.

'He and I had a more intimate relationship than I ever had with Kennedy ...'

Ibid.

'... He's got to cold-bloodedly finish his project and take his chances on the next assignment.'

The Syracuse/NASA Program Project Manager Research Group, interview with Webb. Interviewers: Henry J. Anna, H. George Frederickson and Barry Kelmachter, 15 May 1969.

'Webb ... wanted to have the management of NASA as his final, great undertaking ...'

Courtesy WAMU American University Radio, Washington DC. WAMU 88.5 FM, 'Washington Goes to the Moon', broadcast on 24 May 2001.

'Sit down! ... as far as I'm concerned, Seamans, you're a great big zero.'

Seamans, Robert C. Jr., *Aiming at Targets: The Autobiography of Robert C. Seamans*, pp. 126–8.

'... a gang of NASA people descended on me and swept me into a room and briefed me on what was going on. I was told we were in deep trouble.'

Courtesy WAMU American University Radio, 'Washington Goes to the Moon', 2001.

'... But I didn't share it to quite the extent that Jim believed in it. Maybe he sometimes went a little far.'

NASM Oral History Project, interview with Robert Seamans. Interviewers: Martin Collins and Harry Lambright, 16 December 1988.

Seamans 'didn't believe that clerks within a secretariat would understand the engineering issues sufficiently ...'

Author interview with Seamans, 2005.

'We designed it to a hundred and fifty per cent and it broke at a hundred and forty-four. What the hell do you guys want?'

Gray, Mike, *Angle of Attack: Harrison Storms and the Race to the Moon*, p. 198.

'We were concerned about him ... we weren't sure that he was as thoughtful as he should have been ...'

NASM Oral History Project, interview with Robert Seamans. Interviewers: Martin Collins and Harry Lambright, 16 December 1988.

'Everybody at NASA was courteous, helpful, ... saintly at repeating the same information a hundred times ...'

Mailer, Norman, *A Fire on the Moon*, p. 10.

'More and more, there was a discrepancy between what Webb thought was happening within NASA, and what was actually happening.'

Author interview with Dr John Logsdon, June 2005.

Part Four: Trial by Fire

'Look, we've got to find out what happened, fix it, and be able to fly again. NASA can do that better than anyone else ...'

Webb interview with T.H. Baker, Lyndon Baines Johnson Library, 29 April 1969.

'Anderson let it be known ... that I'd better look out ...'

NASM Oral History Project, interview with Webb. Interviewer: Martin Collins, 15 October 1985.

'Do you really want to kill Apollo?'

From the Earth to the Moon (1997), quoted by permission of Home Box Office (HBO) and Tom Hanks.

'... Here we are, and it's more than thirty years later, but I apparently touched some pretty raw nerves.'
Courtesy WAMU American University Radio, Washington DC. WAMU 88.5 FM, 'Washington Goes to the Moon,' broadcast on 24 May 2001.

'... I don't know why, but he really had it in for us.'
Ibid.

'... They didn't want to discuss it in public, so I pressed them.'
Ibid.

'It has often been said that "People must do what they think is right." In many cases this has been a costly quotation to follow ...'
Quoted from 'An Apollo Report by Thomas Ronald Baron, September 1965–November 1966', NASA History Office.

'... Is there such a thing as the Phillips report?'
All dialogue from the Apollo hearings is taken from formal Congressional records. See: *Investigation into Apollo 204 Accident: Hearings before the Committee on Aeronautical and Space Sciences, United States Senate, February 27, April 17 and May 9, 1967.* See also: *Investigation into Apollo 204 Accident: Hearings before the Subcommittee on NASA, Oversight of the Committee on Science and Astronautics, US House of Representatives, April 10, 11, 12, 17, 21 and May 10, 1967.*

'Webb looked dazed and stunned ...'
Courtesy WAMU American University Radio, 'Washington Goes to the Moon', 2001.

'There's no excuse for volunteering information like that! We're dealing with matters that could result in millions of dollars of lawsuits ...'
Ibid.

'In all the time I worked with Mr Webb ... I can't believe that I didn't keep him informed about everything ...'
Ibid.

'There is little confidence that NAA will meet its commitments within the funds available for the Apollo program.'
Letter from Samuel C. Phillips, Major General, USAF, Apollo Program Director, plus accompanying letter from George E. Mueller, NASA Associate Administrator for Manned Space Flight to J.L. Atwood, President, North American Aviation, Inc., dated 19 December 1965. NASA History Office.

'I guess I was surprised ... when the report began to surface ...'
NASM Oral History Project, interview with Dr George Mueller. Interviewer: Martin Collins, 10 January 1989.

Webb acknowledged that he'd had some discussions with Seamans ... but 'if he told me about it, it was casual and not flagged as something important'.
Robert Sherrod interview notes, 28 April 1971. NASA History Office.

'Mueller was an even stronger character than Brainerd Holmes ... You have to keep in mind that the Phillips Report was originally made for Mueller to see, not Webb.'
Author interview with Dr John Logsdon, June 2005.

'Jim was a person who wanted quick and straightforward answers ... It was a stressful time.'
NASM Oral History Project, interview with Dr George Mueller. Interviewer: Martin Collins, 10 January 1989.

'You must realize, what Webb's saying about George, he's also saying about you.'
NASM Oral History Project, interview with Dr Robert Seamans. Interviewers: Martin Collins and Dr Harry Lambright, 16 December 1988.

'I'm not a psychiatrist but I would say that the fire came as a terrible blow to him.'
Seamans, Robert C. Jr., *Aiming at Targets: The Autobiography of Robert C. Seamans*, p. 144.

'... I mean – he thought this was *portable*. And all of a sudden ... this great management system collapsed.'
Courtesy WAMU American University Radio, 'Washington Goes to the Moon', 2001.

All these unfortunate truths were apparent by the time Floyd Thompson delivered the final draft of NASA's report into the fire on the weekend of 8 April 1967.
'Report of Apollo 204 Review Board', 5 April 1967. NASA History Office.

'It was a continual sort of a rumble of desires ... the NASA people themselves in many cases took quite a lot of time.'
NASM Oral History Project, interview with J. Leland Atwood. Interviewer: Martin Collins, 12 January 1990.

'I think it's rotten and I'm going to blow it out of the water.'
Robert Sherrod interview notes, 2 August 1968. NASA History Office.

'... People are quite willing to accept risk for race drivers ... you don't have a Congressional investigation or a Presidential commission looking at it.'
NASM Oral History Project, interview with Dr George Mueller. Interviewer: Martin Collins, 10 January 1989.

Baron and his wife and two children were killed when a train smashed into his car at a railroad crossing.
'NAA/Apollo Critic Dies in Car Accident', *Space Business Daily*, 8 May 1967. See also: 'Apollo Critic, Wife, Daughter Killed in Crash', *Atlanta Constitution*, 29 April 1967.

'... In a lot of ways, having the sort of job we did humanizes the Russians ... pretty soon you come to know them better than your own people.'
Perry Fellwock, interview with Jamie Doran and Piers Bizony, April 1997.

'I can't face the American people and tell them we're going to pay the contractor bonuses after three people got burned up ...'
Paul Dembling story, quoted in Trento, Joseph, *Prescription for Disaster*, p. 67.

'He was a competent guy ... but I never saw anyone quite so agitated as he became during those hearings.'
Quoted in Gray, Mike, *Angle of Attack: Harrison Storms and the Race to the Moon*, p. 250.

'Jim took all this personally. He became terribly tense. Migraine headaches, which he tended to have anyway, were exacerbated ...'
Seamans, *Aiming at Targets*, p. 144.

Part Five: Dark Side of the Moon

'To persons familiar with NASA's organization, it is utterly inconceivable that Webb would not have known about such a far-reaching document as the Phillips Report.'
'Apollo: A Shining Vision in Trouble', *Sunday Star*, 21 May 1967, p. A-12.

'It can be stated quite without question that this sort of letter simply would not be written ... without the knowledge of the agency's administrator.'
Ibid.

'Do you always tell the truth to Congress?'
Robert Sherrod interview notes, 2 August 1968. NASA History Office.

'He was supposed to get all the information on how to advise people in Washington ... He got us in trouble, and so ... I had to let him go.'
NASM Oral History Project, interview with J. Leland Atwood. Interviewer: Martin Collins, 24 June 1989.

In March 2005, Washington insiders were saddened by the news that a popular socialite had died at the age of 80.
'Socialite Nina Lunn Black Dies', *Washington Post*, 13 March 2005, p. C11.

'... He would tie himself to political people, and more often than not he would use them to his advantage for a job or whatever it might be.'
Interview with Harry Easley, 24 August 1967, by J.R. Fuchs, courtesy of the Harry S. Truman Presidential Library.

... one official did let slip that 'we may be making Mr Black look pretty successful'. The rest of the debt was apparently 'lost in the shuffle'.
'Air Force Silent on Black's Trial', *New York Times*, 23 April 1964, p. 13.

'For a consultant to have to go to somebody who's friendly ... what is freedom of information? I don't know. But it's important.'
NASM Oral History Project, interview with J. Leland Atwood. Interviewer: Martin Collins, 24 June 1989.

Storms and his colleagues on the engineering floor despised Black. 'He would have sold his grandmother if he could make a buck.'
Gray, Mike, *Angle of Attack: Harrison Storms and the Race to the Moon*, p. 118.

'Half a million per year just wasn't enough money for Fred. He was a playboy ...'
Baker, Robert Gene, *Wheeling and Dealing: Confessions of a Capitol Hill Operator*, p. 121.

On the manager's desk a fleshy cherubic face stares out from a framed *Life* magazine cover. 'Bobby Baker's Bombshell', the strapline reads.
The author is grateful to the Carousel's present-day manager Michael James for many fascinating stories.

'Kerr loved Baker. Bobby was his ears, nose, throat, hands and feet in the Senate ...'
Quoted in Young, Hugo, *Journey to Tranquillity*, p. 155.

'He invited me to his office and we played some gin rummy. Then he asked me if I'd like some stock in a bank ...'
Ibid. p. 158.

'Black was sent to me by Senator Kerr on three different occasions. We turned every one of them down.'
Webb interview with T.H. Baker, Lyndon Baines Johnson Library, 29 April 1969.

Black also now owned a modest piece of the very same bank, Farmers and Merchants, where the newly expanded Tulsa plant did its banking.

'Baker Dealings Traced to Kerr', *New York Times*, 31 January 1964, p. 10.

'... He was a Senator and an influential one. This was a pitch and certainly we responded, if you want to look at it that way.'

NASM Oral History Project, interview with J. Leland Atwood. Interviewer: Martin Collins, 24 June 1989.

'... If the Senator was alive, he'd be helping ... I want you to know, North American and Fred Black aren't backing up one inch.'

Courtesy of Stony Brook State University of New York/C. David Heymann Collection, box 16b, item 9: 'Fred B. Black Jr. Surveillance/FBI.' See also: Young, *Journey to Tranquillity*, pp. 154–5.

'All three men have gambling interests in Las Vegas.'

'Baker Dealings Traced to Kerr', *New York Times*, 31 January 1964, p. 10.

'I want to help Bobby. I'll get you the financing if you guys want to go into the vending business. There's a fortune to be made.'

Baker, Robert, *Wheeling and Dealing*, p. 125.

In a $300,000 damage suit, filed against Serv-U in September 1963, Capitol claimed that Baker and Black had 'conspired maliciously' to force them out.

'75 Years of IRS Criminal Investigation History', Department of the Treasury Internal Revenue Service document 7233 (Rev 2-96), catalogue no. 64601H.

He wasn't the only one. Eventually, 125 IRS staffers went to prison and nearly 400 resigned.

'The Death of Conscience', *The Public i: Newsletter of the Center for Public Integrity*, Vol. 5, No. 4, October 1998, pp. 2–3.

'Suit Against Aide Disturbs Senate.'

New York Times, 5 October 1963, p. 44.

Jordan promised that he really was 'trying to keep the Bobby thing from spreading ... it might spread a place where we don't want it to spread ...'

Telephone transcript, 6 December 1963, courtesy of the Lyndon Baines Johnson Library/Michael R. Beschloss.

The following month, Baker collected another $50,000 from grateful campaign contributors. At least $16,000 more followed in November.

'Baker Deals Described By 7 Witnesses', *Chicago Tribune*, 11 January 1967, p. 3.

'He wanted to let me know he loved me and loved my family. He said, "Bob, I hope this is going to be the best Christmas ever for you ..."'

'Baker Rebutted On Big Kerr Loan', *New York Times*, 26 January 1967, p. 1.

'They want to get into campaign expenses, Baker putting out money to senators, controlling what's going on in committees – it's the damnedest mess you ever saw ...'

Telephone transcript, 14 May 1964, courtesy of the LBJ Library/Michael R. Beschloss.

In the end, the jury remained unconvinced and found Black guilty of evading $91,000 in taxes. Sirica sentenced him to fifteen months in jail.

'Black Gets Up To Four Years for Tax Evasion', *New York Times*, 20 June 1964, p. 9.

'... So I don't want one case of corruption. I don't want anything close to it. That means my mother and my brother and my dead daddy ...'
Telephone transcript, 21 April 1964, courtesy of the LBJ Library/Michael R. Beschloss.

Black had by now admitted to at least some of his tax liability from the North American and Serv-U years, amounting to $140,000 ...
'75 Years of IRS Criminal Investigation History', Department of the Treasury Internal Revenue Service document 7233 (Rev 2-96), catalogue no. 64601H.

'... When they got the contract, you had five times as many happy people as you had unhappy people. That's democracy in its finest form.'
Young, *Journey to Tranquillity*, p. 160.

'Now, I have to ask you these questions', he began. 'Were you ever a stockholder in the Serve-U Vending Company?'
Webb interview with T.H. Baker, 1969.

'There was no problem as far as we were concerned. The problem was with people who were looking for something to burn Mr Johnson or Senator Kerr ...'
Ibid.

'... It turned out that Senator Kerr arranged for that very bank to loan money to Bobby Baker in a substantial amount. So everyone tried to draw the association ...'
Ibid.

NASA's curious reticence to supply facts ... brought the credibility of NASA and its top management into sharp question.
See: Apollo 204 Accident: Report of the Committee on Aeronautical and Space Sciences, United States Senate, with Additional Views, 30 January 1968.

'It's beyond imagination that you didn't realize that ... these funds came from narcotics. Businessmen simply don't do business this way'
'Former District Lobbyist Convicted in Drug Case', *Washington Post*, 9 March 1985, p. B1.

Part Six: Recovery

... Seamans could only reply: 'I think they'll feel that I did a satisfactory job.'
Seamans, Robert C. Jr., *Aiming at Targets: the Autobiography of Robert C. Seamans*, p. 147.

'... What kind of so-called "tests" do they do before sending this wreck to us? They'd better get this fixed, and fast! Your name is mud around here until they do.'
Gorn, Michael, *NASA: The Complete Illustrated History*, p. 128.

'Even if I wanted to, I couldn't fire him because he was manager of our successful fast-moving Apollo project, and one of the ablest men in the world ...'
NASM Oral History project, interview with Webb. Interviewer: David DeVorkin, 22 February 1985.

'If a person's shock could be transmitted over the telephone, I'd probably have been shot in the head.'
Murray, Charles and Bly Cox, Catherine, *Apollo: The Race to the Moon*, p. 322.

'If the same guys that had tried to burn him with Bobby Baker and Fred Black and all the rest of it got going again ...'
Webb interview with T.H. Baker, Lyndon Baines Johnson Library, 29 April 1969.

'It was worse than I thought it would be. I hated it. I hated people taking pictures of me when I felt so miserable.'
Kearns, Doris, *Lyndon Johnson and the American Dream,* p. 370.
'As we celebrate the successful accomplishment of our great lunar adventure ... we should give proper credit to James Webb, whose ... skill, vision and drive played a major part in its success ...'
'Letters to the Editor', *New York Times*, 5 August 1969, p. 36.

Part Seven: After Webb

This very generalised essay is based on the author's ten-year charting of NASA and its various programmes as space correspondent for the UK popular science magazine *Focus*. More details about the space station project can be found in Bizony's book *Island in the Sky: Building the International Space Station* (Aurum Press, 1996). See also: NASA Contractor Report 4272, July 1990, *Keeping the Dream Alive: Managing the Space Station Program, 1982–1986*, and *Final Report to the President from the Advisory Committee on the Redesign of the Space Station*, US Government Printing Office, 10 June 1993. See also: Vaughan, Diane, *The Challenger Launch Decision: Risky Technology, Culture and Deviance at NASA*, 1996.
'In the 1970s, when President Nixon killed the Apollo program and ended lunar exploration, NASA said we would do all that again one day ... Now the shuttle is going to be replaced with a CEV capsule just like Apollo.'
Author interview with Robert Zubrin, September 2005.

Selected bibliography

Alexander, Charles C., Grimwood, James M. and Swenson, Loyd S., *This New Ocean: A History of Project Mercury*, NASA SP-4201, 1966
Baker, David, *Spaceflight and Rocketry: A Chronology*, Facts On File, 1996
Baker, Robert Gene (with King, Larry L.), *Wheeling and Dealing: Confessions of a Capitol Hill Operator*, Norton, 1978
Bizony, Piers, *Island in the Sky: Building the International Space Station*, Aurum Press, 1996
Brooks, Courtney G., Grimwood, James M., and Swenson, Loyd S. Jr. *Chariots for Apollo: a History of Manned Lunar Spacecraft*, NASA SP-4205, 1979
Cadbury, Deborah, *Space Race*, Fourth Estate, 2005
Chaikin, Andrew, *A Man on the Moon*, Michael Joseph, 1994
Cooke, Alistair, *Talk About America*, Bodley Head, 1968
Cooke, Hereward Lester and Dean, James D., *Eyewitness to Space*, Harry Abrams Inc., 1970
Dallek, Robert, *John F. Kennedy: An Unfinished Life*, Little, Brown and Co., 2003
Dallek, Robert, *Lyndon B. Johnson: Portrait of a President*, Oxford University Press, 2004
Dyson, George, *Project Orion: the Atomic Spaceship, 1957–1965*, Henry Holt and Company, 2002
Feynman, Richard P., *What Do You Care What Other People Think?*, Norton, 1988
Gatland, Kenneth, *The Illustrated Encyclopedia of Space Technology*, Salamander, 1989
Gorn, Michael, *NASA: The Complete Illustrated History*, Merrell, 2005
Gray, Mike, *Angle of Attack: Harrison Storms and the Race to the Moon*, Penguin, 1992
Grimwood, James M., *Project Mercury: A Chronology*, NASA SP-4001 (MSC Publication HR-1), 1963
Harford, James, *Korolev*, John Wiley and Sons, 1997
Harvey, Brian, *The New Russian Space Programme*, John Wiley and Sons, 1996
Heppenheimer, T.A., *Countdown*, John Wiley and Sons, 1997
Kearns, Doris, *Lyndon Johnson and the American Dream*, André Deutsch, 1978
Killian, James R. Jr., *Sputnik, Scientists and Eisenhower*, MIT Press, 1977
Lambright, Henry W., *Powering Apollo: James E. Webb of NASA*, Johns Hopkins University Press, 1995
Launius, Roger, *NASA: A History of the US Civil Space Program*, Krieger, 1994
Mailer, Norman, *A Fire on the Moon*, Weidenfeld and Nicolson, 1970

McDougall, Walter A., *The Heavens and the Earth: A Political History of the Space Age*, Basic Books, 1985
Murray, Charles and Bly Cox, Catherine, *Apollo: The Race to the Moon*, Secker and Warburg, 1989
Neufeld, Michael J., *The Rocket and the Reich: Peenemünde and the Coming of the Ballistic Missile Era*, Harvard University Press, 1995
Pendle, George, *Strange Angel: The Otherworldly Life of Rocket Scientist John Whiteside Parsons*, Weidenfeld and Nicolson, 2005
Seamans, Robert C. Jr., *Project Apollo: The Tough Decisions*, NASA SP-4106, 1996
Seamans, Robert C. Jr., *Aiming at Targets: the Autobiography of Robert C. Seamans*, NASA SP-2005-4537, 2005
Schefter, James, *The Race: The Definitive Story of America's Battle to Beat Russia to the Moon*, Century, 1999
Shepard, Alan and Slayton, Deke, *Moonshot: The Inside Story of America's Race to the Moon*, Turner Publishing Inc., 1995
Sidey, Hugh, *John F. Kennedy, President*, Atheneum Press, 1964
Smith, Robert W., *The Space Telescope: A Study of NASA, Science, Technology and Politics*, Cambridge University Press, 1989
Stuhlinger, Ernst and Ordway, Frederick I. Jr., *Wernher von Braun: Crusader for Space*, Krieger, 1994
Thomas, Lewin J., and Narayanan, V.K., *Keeping the Dream Alive: Managing the Space Station Program, 1982–1986*, NASA Contractor Report 4272
Trento, Joseph, *Prescription for Disaster*, Harrap, 1987
Vaughan, Diane, *The Challenger Launch Decision: Risky Technology, Culture and Deviance at NASA*, University of Chicago Press, 1996
Webb, James E., *Space Age Management: The Large-Scale Approach*, McGraw-Hill, 1969
Young, Hugo (with Sylcock, Bryan and Dunn, Peter), *Journey to Tranquillity: The History of Man's Assault on the Moon*, Jonathan Cape, 1969

Interview, telephone and meeting transcripts

The John F. Kennedy Presidential Library
W. Henry Lambright
Dr John Logsdon (edited and prepared by Eugene M. Emme)
The Lyndon Baines Johnson Presidential Library
Interviewer: T.H. Baker, 29 April 1969
Special acknowledgements also to Michael R. Beschloss for telephone transcripts related to Lyndon Johnson
The NASA History Office
Robert Sherrod, notes concerning interviews with James Webb on 2 August 1968, 8 June 1969, 16 June 1969 and 17 September 1969, courtesy of the NASA History Office
Smithsonian National Air and Space Museum Oral History Project (incorporating the Glennan-Webb-Seamans Project for Research in Space History)
Interviewers: Martin Collins, Dr David DeVorkin, Ms Linda Ezell, Allan Needell, Dr Joseph Tatarewicz, 15, 22 and 29 March, 10 September, 4 November 1985.

Courtesy of the National Air and Space Museum and the NASA History Office
Stony Brook State University of New York/C. David Heymann Collection
The Syracuse/NASA Program Project Manager Research Group
Interviewers: Henry J. Anna, H. George Frederickson, Barry Kelmachter
Courtesy of the NASA History Office
The Harry S. Truman Presidential Library
Interview with Harry Easely, 24 August 1967, by J.R. Fuchs
WAMU American University Radio, Washington DC, WAMU 88.5 FM
Investigation into Apollo 204 Accident: Hearings before the Committee on Aeronautical and Space Sciences, United States Senate, February 27, April 17 and May 9, 1967
US Government Printing Office, 1967
Investigation into Apollo 204 Accident: Hearings before the Subcommittee on NASA, Oversight of the Committee on Science and Astronautics, US House of Representatives, April 10, 11, 12, 17, 21 and May 10, 1967
US Government Printing Office, 1967

Index

The Science of Doctor Who

Paul Parsons

Have you ever wondered how Daleks climb stairs? How Cybermen make little Cybermen? Or where the toilets are in the Tardis?

Doctor Who arrived on TV screens in 1963. Since then, across light-years and through millennia, the journeys of the Time Lord have shown us alien worlds, strange life-forms, futuristic technology and mind-bending cosmic phenomena. Viewers cowered terrified of Daleks, were amazed with the wonders of time travel, and travelled through black holes into other universes and new dimensions.

The breadth and imagination of the Doctor's adventures have made the show one of science fiction's truly monumental success stories. *BBC Focus* editor Paul Parsons explains the scientific reality behind the fiction.

Discover:

- why time travel isn't ruled out by the laws of physics
- the real K-0 – the robot assistant for space travellers built by NASA
- how genetic engineering is being used to breed Dalek-like designer life-forms
- why before long we could all be regenerating like a Time Lord
- the medical truth about the Doctor's two hearts, and the real creature with five of them.

Hardback £12.99 ISBN 10 1 84046 737 1 ISBN 13 978 1840467 37 6

Issigonis
The Official Biography
Gillian Bardsley

'An authoritative account of the life of this British icon ... a fascinating, beautifully presented study of an enigmatic man.' *Oxford Times*

Sir Alec Issigonis lived his life in parallel with the rise and decline of the British motor industry, becoming its first and only celebrity. He was the designer of the Morris Minor and the Mini, two of the most important British car designs of the 20th century.

Gillian Bardsley, Archivist for the British Motor Industry Heritage Trust, draws on original sources and Issigonis' own privately recorded memories in this first definitive account of the great designer's life.

Beautifully presented, including many of Issigonis' original sketches, as well as photographs of his personal life and his famous cars, this is an absorbing and highly acclaimed account of a gifted but complex man.

Paperback £12.99 ISBN 10 1 84046 778 9 ISBN 13 978 1840467 78 9

The Men Who Measured the Universe

John and Mary Gribbin

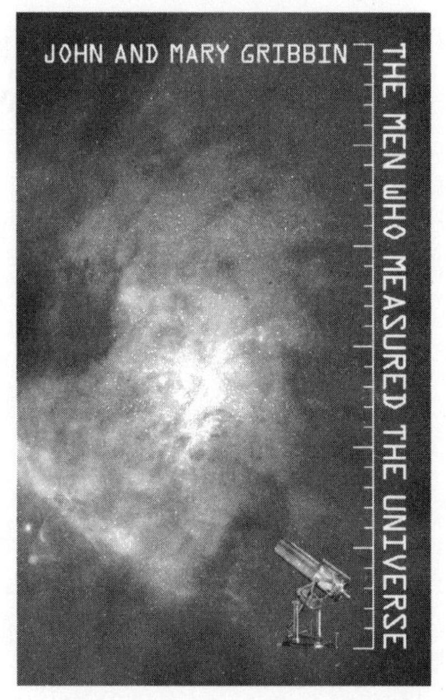

Less than a hundred years ago, astronomers believed that the stars of the Milky Way made up the entire Universe. By the end of the 20th century, they knew that the Milky Way is just one island galaxy among hundreds of billions of light-years of space. With the latest telescopes, we can see 'up' to galaxies 10 billion light-years from Earth.

This revolution in our understanding of the cosmos and our place in it happened within the span of a single human lifetime, through a combination of new technology, in the form of better telescopes, and the dedication of a handful of pioneers obsessed with measuring the distance scale of the Universe. It is their story that is told here, a story that reveals our curious fascination with the night sky, and the hard work, perseverance and spirit of those who sought through their observations to unlock its secrets.

Paperback £6.99 ISBN 10 1 84046 536 0

Why Aren't They Here?
The Question of Life on Other Worlds
Surendra Verma

Is there anybody out there? Are there other life-forms lurking in outer space – or are they already here? Surendra Verma investigates ...

The rate of expansion of our universe is mind-blowing: imagine a pea growing to the size of the Milky Way in less time than it takes to blink. In all this infinite space that we cannot even see, let alone explore, it seems certain that there is some life on other worlds.

Sir Arthur C. Clarke declared that 'the universe is full of intelligent life – it's just been too intelligent to come here'. Journalist Surendra Verma brilliantly outlines the historical, fictional, speculative and emerging scientific opinions on what alien life might be like.

From Aristotle to ET via radio, religion and reincarnation, this fast-moving narrative examines history and dispels myths before focusing on the possibilities lurking in space. In a popular and easy-to-read style, Verma uses current research to speculate on what life is like on other planets, how we might communicate with them, and what Earth might seem like to visitors.

Hardback £9.99 ISBN 10 1 84046 806 8 ISBN 13 978 1840468 06 9

Atom

Piers Bizony

The official tie-in to a brilliant new BBC TV series, *Atom* is the history of the people behind the greatest human scientific discovery ever ...

No one ever expected the atom to be as bizarre, as capricious and as weird as it turned out to be. Its story is one riddled with jealousy, rivalry, missed opportunities and moments of genius.

John Dalton gave us the first picture of the atom in the early 1800s. Almost 100 years later came one of the most important experiments in scientific history, by the young misfit New Zealander, Ernest Rutherford. He showed that the atom consisted mostly of space, and in doing so turned 200 years of classical physics on its head.

It was a brilliant Dane, Niels Bohr, who made the next great leap – into the incredible world of quantum theory. Yet he and a handful of other revolutionary young scientific Turks weren't prepared for the shocks Nature had up her sleeve. Mind-bending discoveries about the atom were destined to upset everything we thought we knew about reality. Even today, as we peer deeper and deeper into the atom, it throws back as many questions at us as answers.

Hardback £12.99 ISBN 10 1 84046 800 9 ISBN 13 978 1840468 00 7

Possessing Genius

The Bizarre Odyssey of Einstein's Brain

Carolyn Abraham

One of Galileo's fingers is in a museum in Florence. Napoleon's severed penis is in the hands, as it were, of an American urologist. And the brain of the greatest thinker of the 20th century lies in two muddy cookie jars under a box behind a beer cooler in Wichita, Kansas.

Follow the bizarre odyssey of Albert Einstein's brain as it roamed the world in mayonnaise jars and courier packages, in car boots and by airmail, and took over one man's life for nearly half a century.

In 1955, Princeton pathologist Thomas Harvey found himself dissecting the corpse of Albert Einstein. He seized the chance to salvage the great thinker's brain, convinced it might unlock the enigma of genius. Harvey became the organ's unlikely custodian – slicing, dicing, pickling and preserving for science and for posterity. *Possessing Genius* is a compelling story about a scientific specimen turned into a holy relic by our rabid culture of celebrity. It is told by science journalist Carolyn Abraham with verve and a uniquely clear insight into the astonishing developments of neuroscience that have driven the extraordinary fate of Einstein's brain.

Hardback £14.99   ISBN 10 1 84046 549 2